FLARE GAS SYSTEMS POCKET HANDBOOK

Gulf Publishing Company
Book Division
Houston, London, Paris, Tokyo

FLARE GAS SYSTEMS POCKET HANDBOOK

K. Banerjee
N. P. Cheremisinoff
P. N. Cheremisinoff

FLARE GAS SYSTEMS POCKET HANDBOOK

Library of Congress Cataloging-in-Publication Data

Banerjee, K.
Flare gas systems pocket handbook.

Bibliography: p.
Includes index.
1. Flare gas systems (Chemical engineering)—Handbooks, manuals, etc. I. Cheremisinoff, Nicholas P. II. Cheremisinoff, Paul N. III. Title.
TP159.F52B36 1985 660.2′83 85-12665

ISBN 0-87201-310-3

ISBN 0-87201-310-3

CONTENTS

4

5

6

PREFACE

Many industries generate significant amounts of waste streams, such as hydrocarbon vapors, which must be disposed of on a continuous or intermittent basis. Examples are off-spec product or bypass streams generated during startup operations. Direct discharge of waste gas streams and vapors into the atmosphere is unacceptable with today's awareness of environmental and safety issues. Neighboring residential areas must be protected against contaminant concentrations exceeding permissible explosion or toxicological threshold limits.

Gas flaring is a standard operation aimed at converting flammable, toxic, and corrosive vapors into environmentally acceptable discharges. This pocket manual provides guidelines for designing and operating conventional flare gas systems. It presents general introductory material to acquaint the engineer with flare gas control techniques, as well as detailed design methodology and formulas. Example problems are included to illustrate design procedures for sizing flare systems.

K. Banerjee
N. P. Cheremisinoff
P. N. Cheremisinoff

NOTATION

A	cross-sectional area
C_p, C_v	specific heats at constant pressure and volume, respectively
C	concentration
d, D	diameter
F	environmental factor
f	friction factor
G	mass flow rate per unit area
g	gravitational acceleration
g_c	conversion factor
H	stack height
h	heating value
K_i	resistance coefficient
L	Length
M	molecular weight
Ma	Mach number
N	line resistance or environmental factor
P	pressure
Q	heat flow rate
q	heat flow per unit area
R	universal gas law constant
R_e	Reynolds number
T	absolute temperature
t	time
U	flare exit velocity
U_w	wind speed
V	velocity
W	vapor mass rate

X distance from flame source
y radial distance
Z compressibility factor

Greek Symbols

ϵ' emmisivity
θ angle
$\varkappa$ ratio of specific heats
μ viscosity
υ specific volume
σ_y, σ_z horizontal and vertical diffusion coefficients

1

OVERVIEW OF FLARING

Introduction

Gas flaring converts flammable, toxic, or corrosive vapors to less objectionable compounds by means of combustion. Flaring is a critical operation in many plants whose design must be based on strict safety principles.

In general, proper planning and layout of process plants require that special consideration be given to the design of various safety facilities to prevent catastrophic equipment failure. These facilities are designed to prevent overpressure and to provide for safe disposal of discharged vapors and liquids. Portions of these facilities are also used as an operational tool for safe disposal of hydrocarbons—particularly during startup and shutdown phases.

Standard pressure-relieving devices most often used are safety and relief valves, rupture disks, pressure-control valves, and equipment blow down valves. Direct discharge of waste or excess vapors to the atmosphere is unacceptable either (1) because of restrictions imposed by local ordinances or plant practices; (2) concentrations of the contaminants at ground or adjacent platform levels may exceed permissible explosion or toxicological threshold limits; and/or (3) meteorological considerations such as severe temperature inversions of long duration may occur, creating hazardous conditions.

Nonhazardous vapors such as waste or low-pressure steam are usually discharged directly to the atmosphere. In contrast, hydrocarbon vapors that are discharged on a continuous or intermittent basis (for example, off-spec product or bypass streams generated during startup) and cannot be directly discharged to the atmosphere should be disposed of through a closed system and burned in a flare.

Types of Flares

There are basically two types of flare systems, namely, elevated flares and ground flares.

In an elevated flare system, combustion reactions are carried out at the top of a pipe or stack where the burner and igniter are located. A ground flare is similarly equipped except that the combustion takes place at or near ground level. Three types of ground flares are in general use:

1. The type that uses a water spray to disperse the combustion gases.
2. The Venturi type that depends on the kinetic energy available in the waste gases to inspirate and mix the proper amount of air with the gases.
3. Multi-jet ground flares where the flow of the waste gas is distributed through many small burners.

Figure 1-1 illustrates a ground flare system. The principal components included in this arrangement are a knock-out drum,

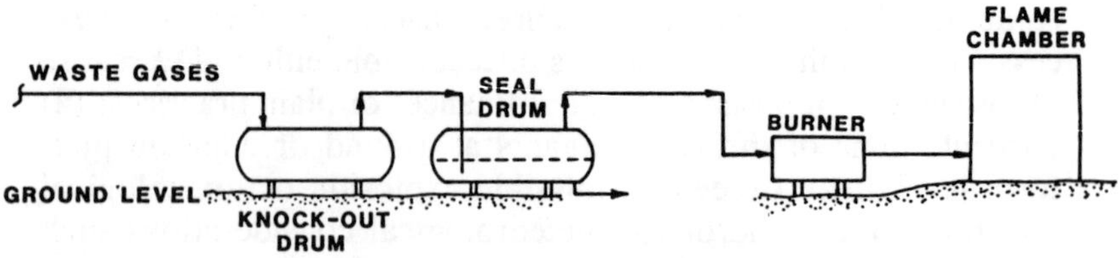

Figure 1-1. Ground flare system.

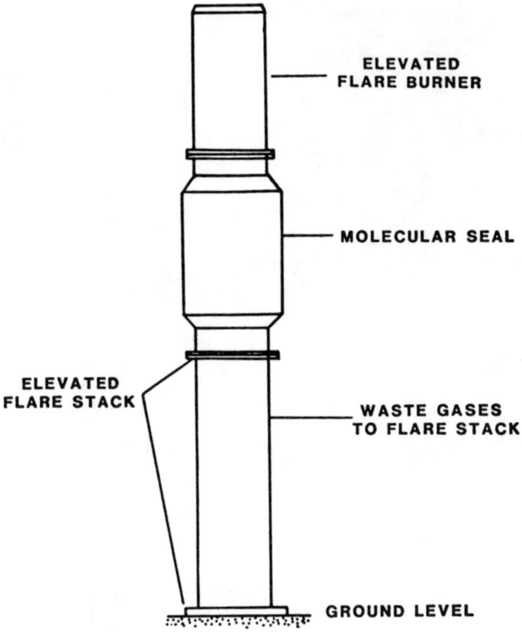

Figure 1-2. Elevated flare system.

multi-jet burners, a refractory lined rectangular flare box, and a seal drum. The flare flame is returned to the inside of the flare chamber.

An elevated flare system is illustrated in Figure 1-2. Relieving gases are sent through an elevated stack from a closed collection system and burned off at the top. The flame generated is open in this case.

Flares are not intended to operate continuously under normal process conditions. However, during plant startups there may be continuous flaring for six or seven consecutive days. Specific industrial applications and flare systems are shown in Figures 1-3 through 1-6.

Figure 1-3. Gasoline vapor flare for loading operation. (Courtesy National Air Oil Burner Co., Philadelphia, PA)

Figure 1-4. An enclosed refinery flare. (Courtesy National Air Oil Burner Co., Philadelphia, PA)

Figure 1-5. Ammonia storage terminal where flare is mounted on tank. (Courtesy National Air Oil Burner Co., Philadelphia, PA)

Figure 1-6. An offshore bridge flare operated in the North Sea. (Courtesy National Air Oil Burner Co., Philadelphia, PA)

Flares are provided with a number of pilot burners at the top. These burn continuously. Also, the flare system is continuously purged with an inert gas such as nitrogen or with natural gas in order to keep the system free of air. During normal plant operations, if continuous venting is required, it may be routed back to a low pressure process stream, to a fuel system, or to an incinerator.

The principal advantages of a ground flare system are:

1. No structural support is required.
2. Erection is relatively straightforward and requires light parts.
3. Maintenance is easy.
4. Operating costs are negligible.
5. The flame of the flare is not visible since it is hidden in a box. It requires less steam to produce a smokeless flare since it produces relatively nonluminous flame because of more controlled combustion at the multiple burner.
6. Finally, with the exception of the Venturi type, it is a fairly quiet system.

One disadvantage of ground flares is that they must be well isolated from the remainder of the plant and property lines, thus requiring considerable space and long interconnecting piping. Concentrations of toxic gases are relatively high because of combustion taking place at ground level. Other disadvantages concern flares that use spray water. These are often avoided because of high water consumption, the possibility of extinguishing the pilot burners, and potential water damage to instrumentation. The Venturi type is almost obsolete because of objectionably high noise levels. The multi-jet type normally used has high initial costs and is capacity-limited.

In contrast, an elevated flare requires less ground area. Because of its high elevation it can be located within a process area or on the periphery of the plant site, since radiation effects and ground-level concentrations of pollutants can be maintained within allowable limits. Piping costs tend to be lower due to smaller and shorter pipe runs. Also, the distance between the

Figure 1-7. Smokeless—no steam flare; flares ethylene with jet mix tips. (Courtesy National Air Oil Burner Co., Philadelphia, PA)

point of discharge from safety valves and the flare stack is less than in the case of ground flares.

A problem with elevated flares is that initial and operating costs are high. Maintenance is also difficult and tedious. The visibility of the flame is the most serious disadvantage and sometimes causes objections from the local community. These systems also require more steam to produce a smokeless flare (see Figure 1-7). A final disadvantage is that noise levels are relatively high.

The selection of the type of flare and special design factors required will be influenced by availability of space, characteristics of the flare gas (i.e., composition, quantity, and pressure level), economics including both initial investment and operating costs, and concern over public relations with the surrounding community.

In general, elevated flares are most often recommended. In spite of the numerous advantages of ground flares, the requirement of the large land area and the associated high initial cost makes it less attractive than elevated systems. However, in some cases, visibility of the flame, depending upon local regulations, could be the determining factor.

There are situations when a ground flare is used in conjunction with a second conventional flare, which may be an elevated system. The ground flare is designed to handle the normal flaring requirement. In the event of a major failure, excess flow is automatically diverted by a seal to a second flare. Since the possibility of a major failure is rather remote, it may not conflict with pollution or local site regulations.

Flare System Components

A typical flare system is comprised of the following components:

1. Relief, safety, and depressuring valves (major features of a safety relief valve are shown in Figure 1-8).
2. Pressure-relieving header(s) that convey discharges from safety and pressure-control valves in the process unit to the flare.
3. Knock-out drum located before the flare stack in order to separate any condensate or liquid from the relieving vapors (It is hazardous to burn liquid droplets).
4. Flare stack consisting of the riser structure, molecular seal, and burner tip.

A brief description of the major components follows.

Riser Structure: This normally consists of two or more sections. The flare header enters at the bottom section, which serves as a flare stack knock-out drum where any condensate carried over from the knock-out drum is collected.

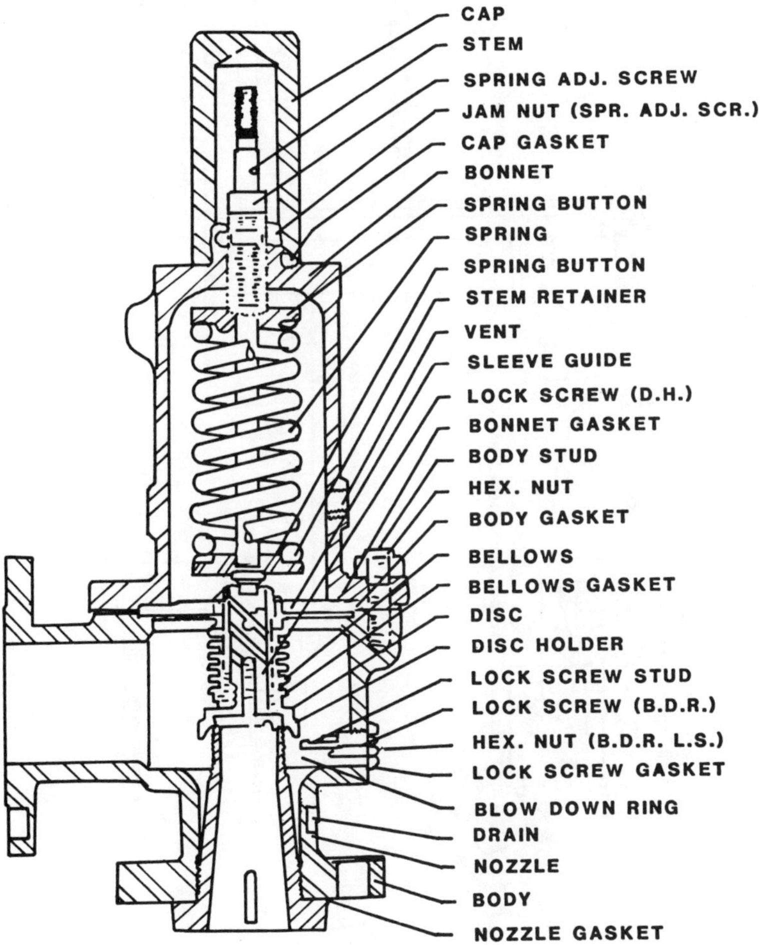

Figure 1-8. Features of a balanced bellows safety relief valve.

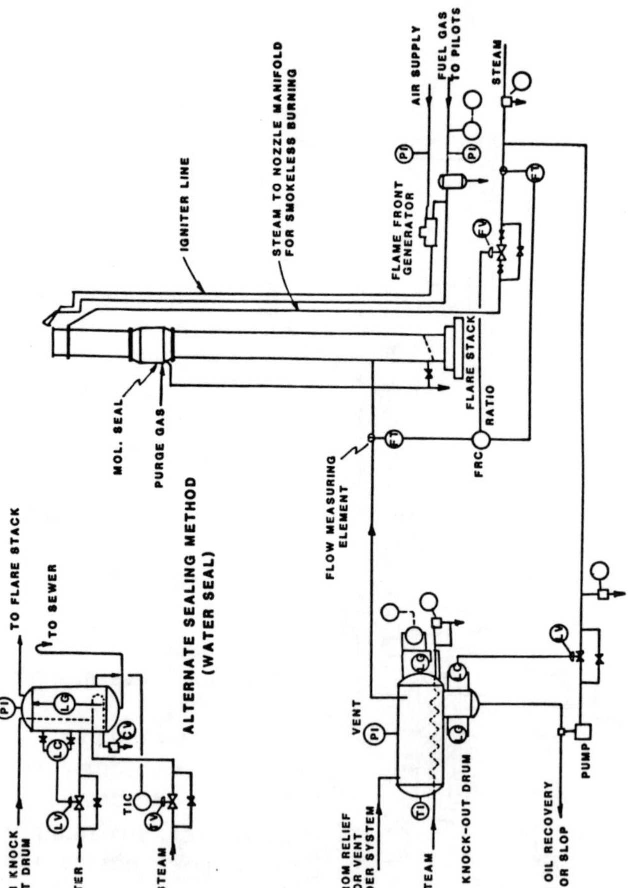

Figure 1-9. Schematic of typical flare gas system (from API, RP 521).

Molecular Seal: This is welded to the riser section. It provides a seal against entrance of air into the flare stack and minimizes the possibility of an explosive mixture forming in the flare system. Briefly, it resembles a bubble cap and creates a seal by using the buoyancy of the purge gas to create a zone where the pressure is greater than atmospheric pressure.

Flare Burner Tip: The burner tip is a complete assembly sealed to the molecular seal outlet. Accessories on the burner tips include about three or four gas pilots, a similar number of pilot gas/air mixture assemblies, and steam supply nozzles for steam injection.

Figure 1-9 shows a schematic diagram of the entire flare system.

The relieving gases from safety-relief valves are collected in a horizontal or vertical knock-out drum through a flare main header. Any condensate carried over along with the gases is knocked down here. A constant liquid level is maintained in the boot. The liquid is pumped to a slop tank or reused in oil recovery facilities. Steam is used for winterizing to prevent freezing. The gas from the knock-out drum is then sent to an elevated flare stack. At the bottom of the stack a liquid seal is maintained. Alternately another seal may be located between the knock-out drum and the flare stack. A positive water seal is maintained by controlling the level. It is also provided with steam for winterizing.

The stack is comprised of a riser section, molecular seal, and burner tip. At the top of the burner tip, pilot burners, which are automatically lighted from a remote place through the igniter line, are positioned. The steam connection is also provided for smokeless flares and a purge gas connection is provided for maintaining an air-free system and to prevent flashback by maintaining pressure at the molecular seal higher than atmospheric pressure. This arrangement prevents air from reentering the stack from the ambient surroundings.

Smokeless Flares

A flame is referred to as being luminous when incandescent carbon particles are present in it. When these particles cool, they form smoke. Several theories have been presented to explain the formation of smoke, but none has been fully established. However, it has been observed that smoke formation mainly occurs in fuel-rich systems where a low hydrogen-atom concentration suppresses the smoke.

Prevention of smoke in flares is normally accomplished in three different ways:

1. By the addition of steam.
2. By making a pre-mix of fuel and air before combustion so as to provide sufficient oxygen for efficient combustion (which is always done in fired boilers and furnaces).
3. By distribution of the flow of raw gases through a number of small burners.

Among these methods, the addition of steam is most commonly used to produce a smokeless flare for economy and superior performance. In steam addition, the raw gas is preheated before it enters the combustion zone of the flame. If the temperature is high enough, cracking of the hydrocarbons may occur. This produces free hydrogen and carbon. When the cracked hydrocarbons travel to the combustion zone, hydrogen reacts much faster than carbon. Unless the carbon particles are burned away, they cool down and form smoke. Consequently, in order to prevent smoke, either the hydrogen atom concentration must be decreased to ensure uniform burning of both hydrogen and carbon or enough oxygen must be provided for complete combustion.

There are several theories as to the chemistry of smokeless flares using steam. One of them assumes that the steam separates the hydrocarbon molecules, thereby minimizing polymerization reactions, and forms oxygen compounds that burn at a reduced rate and temperature so as to prevent cracking.

Another theory claims that steam reacts with carbon particles forming carbon monoxide, carbon dioxide, and hydrogen, thereby removing the carbon which forms smoke after cooling.

We shall close this chapter with some common flare system terminology.

Terminology

Relief Valve. An automatic pressure-relieving device actuated by the static pressure upstream of the valve, and which opens in proportion to the increase in pressure. It is used primarily for liquid service.

Safety Valve. An automatic pressure-relieving device actuated by the static pressure upstream of the valve and characterized by a rapid full opening or pop action. It is used for gas or vapor service. In the petroleum industry it is normally used for steam and air.

Safety-Relief Valve. An automatic pressure-relieving device suitable for use as either a safety valve or relief valve, depending on application. In the petroleum industry, it is normally used in gas and vapor service or for liquid.

Pressure-Relief Valve. A generic term applying to relief valves, safety valves, or safety-relief valves.

Rupture Disk. Consists of a thin metal diaphragm held between flanges, which bursts open under pressure and must be replaced before restarting the system.

Maximum Allowable Working Pressure. The highest pressure in the equipment at which the primary pressure relief valve is set to open. It depends upon the type of material, the thickness of the pressure vessel and the service condition.

Operating Pressure. The pressure at which a vessel normally operates. A vessel is usually designed for maximum allowable working pressure. Thus, there should be a suitable margin between the normal operating pressure and the maximum allowable working pressure in order to prevent undesirable operation of the vessel by frequent popping up of the relief device.

Set Pressure. The pressure at which the relief device is set to open. It is normally less than or equal to the maximum allowable working pressure.

Accumulation. A pressure increase over the maximum allowable working pressure when a relief device is discharged. It is expressed as a percentage of the maximum allowable working pressure.

Back Pressure. Pressure on the discharge side of a safety-relief valve. This is of two types:

1. *Super-imposed back pressure:* It is the pressure in the discharge header of a safety-relief valve before it discharges. Thus, if a valve discharges to the atmosphere, back pressure is atmospheric pressure.

2. *Built-up back pressure:* It is the back pressure developed as a result of flow after the safety relief valve opens. In the preceding example, this would be greater than atmospheric pressure.

2

DESIGN PRINCIPLES

Introduction

The design of a flare system first requires a detailed analysis of the possible situations that can cause discharge from pressure-relief valves, thus establishing the maximum loading for emergency operations. The maximum load of a system is comprised of the individual loads contributed by the entire process. That is, a conservative design is one that assumes all contributors for the process are relieving simultaneously under any emergency condition.

From a practical standpoint, it is preferable that relieving of overpressures to the flare systems via the pressure-relief valves should be kept to a minimum, since after "popping," the pressure-relief valves often do not reseat. This results in leakage, and consequently reduces the recovery of valuable products. For minor operational upsets, and especially at startup, where flaring for periods of time is often necessary, the use of pressure-relief valves is undesirable. Therefore, overriding pressure-control valves strategically located on some equipment are often used in addition to the pressure-relief valves. Examples of such locations are the suction sides of compressors, overhead product lines of fractionating columns, at the beginning or at the end of a series of high-pressure reactors, etc. The set point of the pressure-control valve is kept above the operating pressure but below the set-point of the pressure-relief valve. They are sized to handle about 40 to 100 percent of the flow of the safety valves. In about 90 percent of the emergency situations, the system is depressurized through these valves to keep the operation of the pressure-relief valves to a minimum.

Besides overriding pressure-control valves, other remotely controlled valves such as motor-operated or solenoid-operated valves are also used for depressurizing the system in emergency situations—especially during a fire.

Causes of Overpressure

Pressure vessels, heat exchangers, operating equipment, and piping are designed to contain the system pressure. The design is based upon the maximum pressure anticipated during operation of a piece of equipment and the corresponding temperature. This pressure is known as the maximum allowable working pressure and is about 10% higher than the normal operating pressure. Pressure relief valves are at or below the maximum allowable working pressure in order to protect the vessel. The relieving rate of the pressure-relief valves depends upon the cause of system overpressure.[1]

The principal causes of system overpressure are classified as *operational failures* and *plant fires.*

Operational Failures

Typical operational failures are caused by closed outlets on vessels, inadvertent valve openings, utility failures, and a variety of miscellaneous actions.

Closed Outlets on Vessels. Inadvertent closure of a block valve at the outlet of a vessel while the plant is on stream may expose the vessel to a pressure that exceeds the maximum allowable pressure. A pressure-relief valve is required to protect the vessel under this situation. If two vessels are in series and isolated by individual block valves, then each vessel must be protected by an individual pressure-relief valve. Omission of the block valve in between the vessels or locking the same valve in the open position will make it a common system. This arrangement may be protected by a single pressure-relief valve. As in the case of block valves, every control valve should be considered as subject to inadvertent operation causing overpressure in the upstream section.

Inadvertent Valve Opening. Inadvertent opening of any valve from a higher pressure source, such as high-pressure steam or process fluids connected to a low-pressure system—causes an overpressure exceeding the maximum allowable working pressure. A pressure-relief valve is required to protect the system. However, if the block valve is intended for isolation only and normally remains closed, a pressure-relief valve may be avoided by locking or sealing closed the same block valve.

Utility Failure. Loss of any utility, whether plantwide or local—may cause overpressure. This is a serious issue since in most cases utility failure determines the controlling flare load. Common utility services that may fail along with the equipment affected are listed here:

Utility Failure	Equipment Affected
Electric power	1. Pumps for circulating cooling water, boiler feed, quench, or reflux. 2. Fans for air-cooled exchangers or cooling towers. 3. Compressors for process vapor, instrument air, vacuum or refrigeration. 4. Instrumentation 5. Motor operated valves
Cooling Water	1. Condensers for process or utility services. 2. Coolers for process fluids, lubricating oil or seal oil for pumps or compressors.

	3. Jackets on rotating or reciprocating equipment.
Instrument Air	1. Transmitters and controllers. 2. Process regulating valves. 3. Alarms and shutdown systems.
Steam	1. Turbine drives for pumps, compressors, or electronic generators. 2. Reboilers. 3. Reciprocating pumps. 4. Process using direct steam injection. 5. Eductors.
Fuel	1. Boilers for process steam. 2. Reheaters (reboilers). 3. Engine drivers for pumps or electric generators. 4. Process heaters. 5. Utility boilers. 6. Gas turbines.

Miscellaneous—Examples of miscellaneous actions/conditions of operation failures are as follows:

Reflux Failure of a Fractionating Column—Reflux failure causes flooding of condensers, resulting in an overpressure that might exceed the maximum allowable working pressure.

Heat Exchanger Tube Failure—When the tube side pressure of an exchanger is much lower than the shell side, rupture of a tube will overpressure the tube side. It should be protected with a pressure-relief valve.

Internal Explosion—These are not predictable for conventional refinery installations. For some chemical reactions, it may be possible to predict the possibility of an explosion, in which case, special rupture disks are installed for quick disposal of vapors.

Chemical Reaction—Chemical reaction vessels may be overpressured because of an unbalanced reaction, especially if it is exothermic. Normally, very sophisticated controls are used along with a safety valve in these cases.

Hydraulic Expansion—When a cold fluid is blocked in on hot exchanger surfaces, it will expand and cause a rise in pressure.

Accumulation of Noncondensables—Noncondensables do not accumulate under normal conditions since they are released with the process streams; however, with certain piping configurations, it is possible for noncondensables to accumulate to the point that they may prevent condensation of a process stream in a condenser, thereby causing overpressure.

Failure Because of Plant Fire

Any process equipment in an operating plant that handles or processes flammable liquid or gases may be exposed to fire at some time during its operation. If an open, free-burning fire occurs, heat will be absorbed by the vessel or other equipment exposed to flames. This is true even for vessels containing fluids that are not flammable. If this heat absorption continues, pressure will develop inside the vessel by vapor generation or by expansion of the fluid. A pressure-relief valve is required to protect the vessel and relieve the generated vapor.

Estimating Relief Rates from Pressure: Relief Valves or Other Safety Devices

Each pressure-relief valve should be individually analyzed for any probable causes of overpressure due to operational failure and plant fire.

The valve should be sized for the case that will require the maximum relieving rate. If a fire condition is controlling, two separate safety valves, one for fire condition and the other for operational failure, should be provided since the fire situation is less likely to occur.

Guidelines for determining individual relieving rates are as follows:

For Operational Failures

These are best explained through examples of different operating units. For instance, consider a fractionating column where different causes of overpressure may be analyzed as follows:

Blocked outlet condition of the overhead vapor line by inadvertent closure: In this case, it may be assumed that heat input to the reboiler is normal, and the reflux will still be maintained since the accumulation for the overhead condenser has the holding capacity for about 10 minutes. Hence, the relieving capacity of the pressure-relief valve may be assumed as the normal vapor load to the overhead condenser. The relieving pressure will be the set pressure of the pressure-relief valve and the temperature will be the boiling point corresponding to that pressure. Credit may be taken for the vaporization rate because of increases in the boiling point with an increase in pressure from normal operating pressure of the column to the set pressure of the valve.

Cooling water failure: The cooling water pump may trip because of power failure or some other operational problem. Under this situation, the overhead vapor will not condense in the condenser and because of the vapor accumulation, the pressure will rise. The reflux can still be maintained for about 10 minutes because of the holding capacity of the accumulator. The relieving capacity of the pressure relief valve will also be the normal vapor rate to the condenser. In this case also, credit may be taken for the rate of vaporization because of the increase in pressure.

Reflux failure: This may be associated with the malfunction of the reflux control valve, or instrument air failure, or any other operational problem. In this case, the overhead condenser becomes flooded with condensate. As a result of this, overhead vapor cannot condense and pressure starts to build up. Also, the relieving rate will be the same here as well as in the conditions cited in the preceding paragraph.

If a fractionating column is provided with a side-stream reflux, its failure will also result in condenser flooding and the relieving rate will be the same as just described.

Abnormal heat input by reboilers. If the temperature controller at the reboiler fails because of instrument air failure or for any other reason, the rate of vaporization may increase. If the vaporization rate exceeds the rate of vapor condensation, the pressure will build up. In this case, the relieving rate should be the difference between the maximum rate of overhead vapor and the maximum rate of condensation of the condenser. In the absence of data, the relieving rate may be assumed to be the normal vapor load to the condenser.

Other examples of units subject to operational failure are heat exchangers, pumps, compressors, heaters, reactors, etc. Only those items that handle vapors will be mentioned here since liquids do not discharge into the flare headers.

An exchanger or a compressor may be pressurized because of a blocked outlet condition. The pressure-relief valve provided should be designed so that its operation corresponds to the normal flow rate of the exchanger (shell or tube, whichever is blocked) or the compressor.

Compressors are provided with intercoolers and aftercoolers and these exchangers normally use cooling water as the cooling medium. In case of cooling water failure or failure of a temperature controller, the temperature of the gas being compressed may be so high as to exceed the allowable temperature of the compressor loop.

A high temperature condition normally triggers automatic shutdown of the compressors. When the compressor shuts down while the gas stream is still flowing, the suction side pressure-relief valve blows out. In an actual operation, there usually is an overriding pressure-control valve set at a pressure lower than the pressure-relief valve, and which opens first. In these cases, the relieving rate should be assumed to be the normal flow of the compressor.

The relieving rate corresponding to the normal or unbalanced reaction rate in a reactor can be approximated from the kinetic data.

For Relief Loads Due to Fire

Wetted surface area. The surface area of a vessel exposed to fire, and which is effective in generating vapor, is that area wetted by its internal liquid level up to a maximum height limitation of 25 ft above grade, which is the normal practice based upon the flame length. "Grade" is defined as any horizontal solid surface on which liquid could accumulate, i.e., roofs, solid platform, etc. The contents under variable level conditions would ordinarily be taken at the average inventory. Liquid-full vessels, horizontal or vertical (such as treaters), operate with no vapor space, and the wetted surface would be the total vessel area within a height limitation of 25 ft above grade. It should be noted that, in such a vessel, at the start of a fire the opening of the pressure relief may be due to thermal expansion of the liquid. However, the pressure-relief valve should be sized based upon the vapor generator at the relief pressure and the boiling point corresponding to that pressure.

The surface areas of typical vessels used in process operations are as follows:

- *Surge and reflux drums:* The wetted surface should be calculated using the high liquid level or 50 percent of the total vessel surface, whichever is greater, since 50 percent is the normal liquid level in these vessels.
- *Knockout drums:* Knockout drums usually operate with only a small amount of liquid at the bottom of the drum. If the normal liquid level is not known, the level at the high level alarm should be used to estimate the wetted surface.
- *Fractionating columns:* Fractionating columns usually operate with a normal liquid level in the bottom of the column plus a level on each tray. However, the entire wall of a frac-

tionating column within a fire height limitation of 25 ft should be considered as wetted.

- *Working storage tanks:* Here the liquid level is independent of operation, and, therefore, the maximum liquid level should be used for determining the wetted surface. The wetted surfaces of spheres and spheroids are calculated as the area of the bottom half of the vessel or up to a height of 25 ft, whichever gives the greater surface area.

Heat absorbed by vessels. Where suitable drainage is provided to preclude an accumulation of flammable liquids directly under a vessel, the total heat input rate to the vessel may be computed as follows:

$$Q = 21{,}000\ FA^{0.82} \tag{2-1}$$

where

Q = total heat absorbed in BTU/hr
A = wetted surface in sq ft
F = environment factor

This equation is recommended by the American Petroleum Institute, RP520.

Using the appropriate value of the wetted surface and the value of factor "F" tabulated below for different thicknesses of insulation, the heat input rate may be calculated.

F = 1.0 for bare surface
F = 0.3 for 1 in. thickness of insulation
F = 0.15 for 2 in. insulation
F = 0.075 for 4 in. insulation

If insulation is required but the thickness is not known, an F-value of 0.3 is recommended. The amount of heat absorbed can also be obtained from Figure 2-1.

If drainage is not provided for the area under the vessel (i.e., diked or curbed areas around a tank), then vapor relief for fire

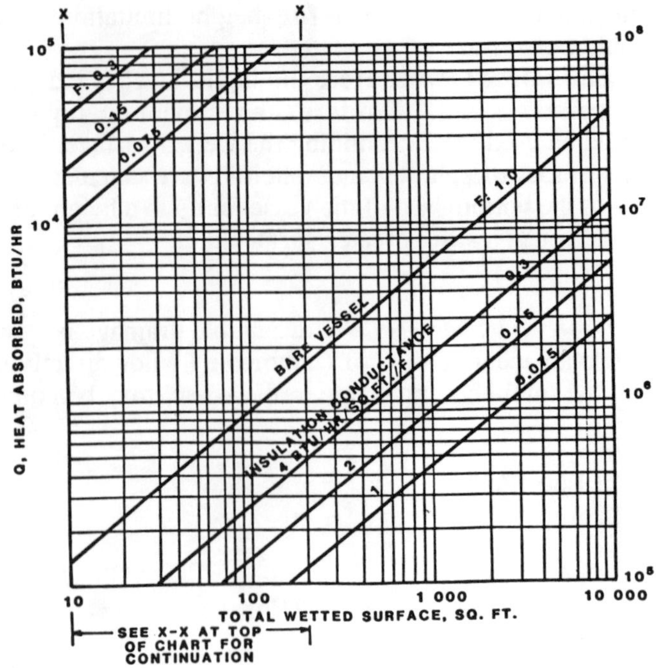

Figure 2-1. Graphical presentation for formula for heat absorption calculation.

exposure should be computed using the following heat input criteria:

20,000 BTU/hr/ft^2 for an uninsulated vessel
10,000 BTU/hr/ft^2 for 1 in. insulation
6,000 BTU/hr/ft^2 for 2 in. insulation
3,000 BTU/hr/ft^2 for 4 in. insulation

These values are based on the wetted surface up to the normal liquid level, provided the insulation is fireproofed. If insulation is not fireproofed, the vessel should be assumed as bare.

Vapor generated. For a fluid below the critical point (i.e., at relieving temperature and pressure), the rate of vapor released is:

$$W = \frac{Q}{\lambda} \tag{2-2}$$

W = vapor release rate lbs/hr
Q = total heat input BTU/hr
λ = latent heat of fluid in vessel, evaluated at the relief valve inlet pressure BTU/lb

No credit is normally taken for the sensible heat capacity of the fluid in the tank.

For a fluid above the critical point, i.e., when pressure-relief conditions are near or above the critical point, the rate of vapor discharge depends upon the rate at which the fluid will expand as a result of the heat input. The latent heat of vaporization at or near the critical point is almost zero in this case.

Maximum Vapor Load to be Flared

After relieving loads of individual pressure-relief valves have been calculated, a detailed study is required to determine how these relieving situations are related to each other. The simultaneous occurrence of two or more contingencies (known as double jeopardy) is so unlikely that this situation is not usually considered as a basis for determining the maximum system load. In determining the maximum load from a single contingency, all directly related contingencies that influence the load must be considered. For example, in a plant where a single boiler or source of steam is used for both process drives and electric power generation, a failure of a steam source (a single contingency) can cause simultaneous loss of power failure (a directly related contingency). If the electrical system had an alternate source of supply, then only the loss of steam would be considered, provided the elapsed time for transfer switching was so long as to be ineffective. In this situation, power failure would not be a contingency directly related to the loss of steam.

If a certain contingency were to involve more than one unit, then all of them must be considered in their entirety. For example, if there is more than one reaction vessel in series, then in the event of a runaway reaction, all the reactors will be protected by a single pressure-relief valve. The same situation can occur in the case of two or more fractionating columns in series. Again, in the case of multi-stage compressors having individual pressure-relief valves at each stage, the relieving rates from the pressure-relief valves are not additive.

Another example of a directly related contingency would be failure of coolant to one unit when the coolant is supplied from a second unit. This is shown in Figure 2-2. If the compressor trips out, safety valve No. 2 blows out since the pressure at the flash drum builds up because of high temperatures. Simultaneously, safety valve No. 1 also blows because of coolant failure (either total failure or partial failure depending upon the heat input and condensing temperature of the process vapor in the condenser).

Since double jeopardy is not usually considered, the maximum load can be based upon any one of the following contingencies:

1. Electrical power failure
2. Cooling water failure
3. Steam failure
4. Instrument air failure

In addition, other contingencies may be simultaneous. Again referring to Figure 2-2, coolant failure is related to compressor failure.

Complete failure of electrical power, cooling water, or steam to an entire plant should be studied. Since the odds against complete failure of power are rather astronomical, failure of one bus bar is normally considered, provided the power source is very reliable and backed up by an emergency power supply. Again,

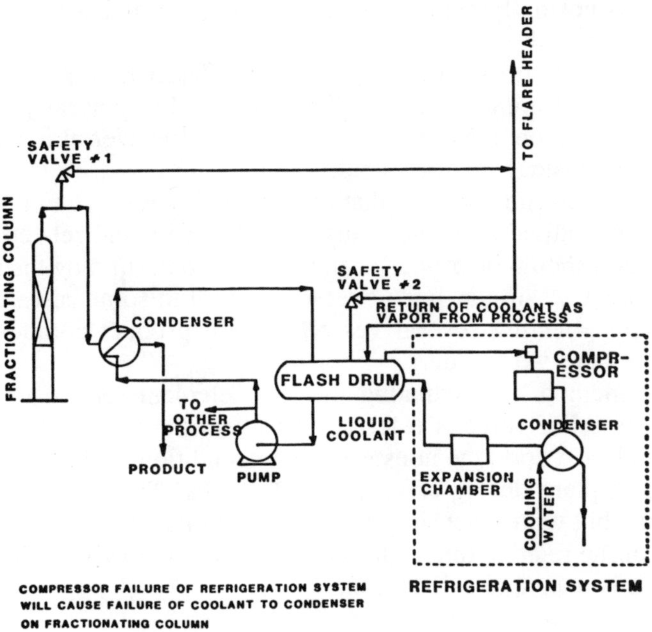

Figure 2-2. Example of two directly related contingencies. (The compressor failure of the refrigeration system will cause a failure of coolant to the condenser of the fractionating column).

depending upon the size of the plant and the degree of the power source reliability, failure of the entire electrical distribution or even the entire main line could be the design basis.

The most common basis for analysis of cooling water and steam failure is the failure of one lateral which means about 50 percent of the total supply. However, entire cooling water failure is also a very common basis of analyzing failure in a small plant. It is common practice to provide the steam generation unit with a back-up U.P.S. unit (Undisturbed Power Supply). Instrument air failure is normally plantwide unless automatic make-up or sparing is provided.

A cause of fire is normally localized. The whole plant is divided into different fire zones. The flare load is generally calculated based upon one or two related zones. However, it is not unusual to consider the total load.

Another consideration is that the time delay relative to the discharge of individual valves caused by the same and related contingencies should be properly studied when determining the maximum load. A similar line of reasoning will in some cases apply to a fire affecting several vessels, where product composition and pressure vary widely.

The method of calculating the time element related to each pressure-relief valve is referred to as Transient Load Analysis. This is based upon the nonsteady state condition in the flare system of a plant during emergency situations. This calculation is tedious, but with some simplified assumptions it provides an estimate of the relative time delays of the individual valves. The following is a sample calculation.

Example

A fractionating column has a safety valve set at 296 psia. Normal pressure of the overhead vapor is 260 psia, Temperature = 100°F. Total system volume = 55,000 ft. Relieving rate estimated as normal vapor rate to the condenser is 260 lbs/sec. Determine the time lag between the beginning of cooling water failure of the condenser and the actual blowing of the safety valve.

Solution

ρ_L = 30 lbs/cft; 16 ft dia.; 25 ft high
No. of trays = 150, level of liquid = 2 in. each tray
MW = 42

Assuming temperature rise = 10°F

$$\rho_1 \text{ (initial density of vapor)} = \frac{42 \times 260}{10.7 \times 560} = 1.82 \text{ lbs/cft}$$

$$\rho_2 \text{ (final)} = \frac{42 \times 296}{10.7 \times 570} = 2.07 \text{ lbs/cft}$$

$\Delta\rho = 0.25$ lb/cu. ft

inventory = 0.25 × 50,000 = 12,500 lbs

volume of vapor space = 50,000 cft

volume of liquid in the tower

  volume of liquid = 5,000 cft

  weight of liquid = 150,000 lbs

Volume occupied by vapor = 55,000 − 5,000 = 50,000 cft

Liquid to be heated from 100°F to about 110°F:

  $\Delta t = 10°F$; $\Delta H = 5.5$ Btu/lb

Total BTU to heat up liquid = 150,000 × 5.5
= 830,000 BTU

Normal heat load to reboiler = 120×10^6 BTU/hr
= 33,000 BTU/sec

Neglecting heat required to heat up metal temperature, time lag:

1. Time required to heat liquid

$$\frac{830{,}000}{33{,}000} = 25 \text{ sec}$$

2. Time required to raise pressure

$$\frac{12{,}500}{260} = 48 \text{ sec}$$

Relieving rate = 260 lbs/sec

Hence, total time lag is 73 seconds.

Table 2-1
Typical Breakdown of Loads from Different Process Areas*

Process Area	Safety Valve (SV) or Process Control Valve (PC)	Description of the Unit Where the Valve is Located	Set Pt PSIG	Relieving Rate lbs/hr	Types of Failure C/W = Cooling Water P = Power, Com = Compressor O = Others, F = Fire Failure	Remarks
100	SV-101	Rerun Tower	30	4,700	C/W or P	
				8,000	F	Valve Sized for Fire
	SV-102	Feed Vaporizer	140	180,000	O	Individual Relief Does not Add to Other Loads
200	SV-202 or PC-202	First Suction Drum of Compressor	40	*245,000	P or C/W	
				4,000	F	
	SV-203 or PC-203	Second Suction Drum of Compressor	70	27,800	O	Individual Relief
				4,000	F	
	SV-204	Third Discharge Drum	270	253,100	O	Individual Relief
				4,000	F	
	SV-215	Dryer Feed Separator No. 1	140	180,000	O	Individual Relief
		Dryer Feed Separator	580	237,000	O	Individual Relief
				10,000	F	

300	SV-301	C_2 Recycle Vaporizer	140	66,550	O	Individual Relief
400	SV-401	Depropanizer	175	65,000	C/W or P	
	SV-402	Methane Stripper	350	8,800	C/W	
500	SV-501	C_3 Compressor Suction	110	100,000	Com	
		C_3 Compressor Discharge	290	680,000	C/W	
600	SV-601 or PC-601	C_2 Compressor Suction	80	120,000	Com	
			TOTAL LOAD	999,500	C/W	
				314,700	P	
				220,000	Com	
				30,000	F	Max. Individual load due to other failure (e.g. blocked outlet, control valve failure, etc.)
				253,000		

**By transient load analysis this load discharges to the flare header after sufficient time (and should be subtracted from total load)*

In large chemical plants or refineries provided with a very reliable power supply, cooling water failure mostly turns out to be the single contingency that determines the maximum flare load.

In a large olefin plant producing mainly ethylene and propylene, the following major contingencies are normally considered:

1. Cooling water failure
2. Power failure
3. Failure of propylene refrigeration compressor

Out of these three contingencies, cooling water failure dictates the controlling flare load. Table 2-1 gives a breakdown of the individual loads from safety valves, types of failure, and the cumulative loads based upon different process areas. The process areas are normally given area numbers such as 100 area, 200 area, etc., in order to identify on the plot plan layout what type of process equipment is involved. As shown by the table, the nominal load that the flare burner will handle is 800,000 lbs/hr and the maximum instantaneous load or the design load is 1,000,000 lbs/hr.

3

DESIGN OF COLLECTION SYSTEMS

Introduction

The relieving vapors from different pressure-relief valves and depressuring valves must first be collected in individual flare subheaders located near each process area. Subheaders must be interconnected to a main flare header which leads to a knock-out drum. Condensates carried over by vapors are separated in this vessel. Vapors leaving the knock-out drum from the top move up the flare stack where they are subsequently burned at the tip.

The number of main flare headers and the individual subheaders connected to them depends upon the type of vapors handled, temperature, and the back pressure limitation of the pressure relief valves. This chapter summarizes design criteria for specifying collection systems arrangements.

Main Flare Header and Subheader Pressure Levels

The pressure level of the flare headers depends on the type of pressure-relief valves used to protect the equipment and the pressure levels of the equipment connected to the flare system.

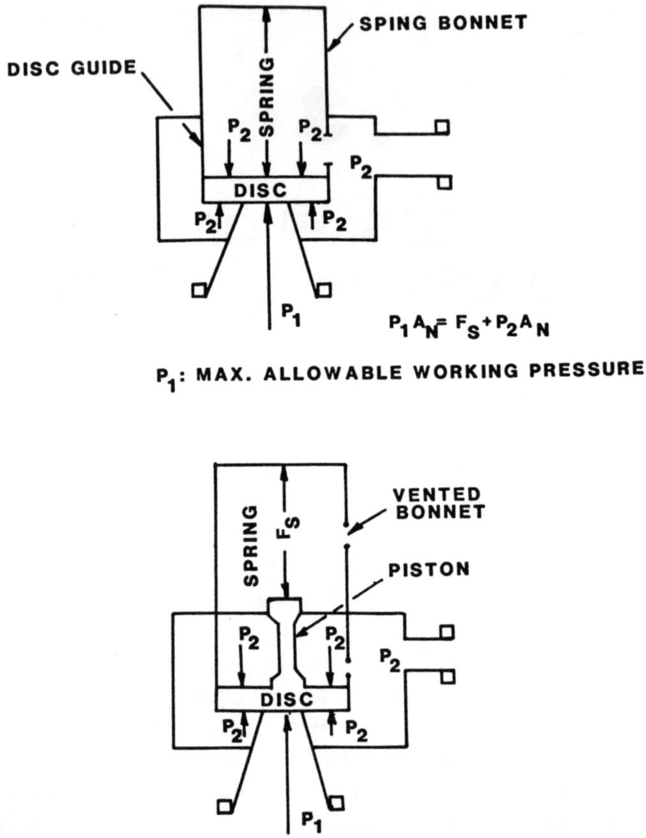

Figure 3-1. Effect of back pressure on (A) conventional safety valve and (B) piston type valve.

The principal types of pressure-relief valves are:

1. Conventional
2. Balanced bellow type
3. Piston type
4. Pilot operated

Conventional pressure-relief valves are those where the disk of the valve is held tight against the inlet nozzle by means of a spring. Figure 3-1A shows a schematic diagram of a conventional safety-relief valve normally used in refineries and chemical plants. This type of valve is least expensive but it is limited by a maximum back pressure of 10 percent of the maximum allowable working pressure. The reason for the limited back pressure is as follows:

Let P_1 be the maximum allowable working pressure of the vessel to which the safety valve is connected. Since this is also normally the set pressure of the safety valve, the spring is so loaded that the total downward force on the valve disk is exactly equal to the total upward force on it exerted by the process vapor when it reaches the pressure equal to P_1. That is, the spring force F_s is related to the inlet nozzle area A_N through applied pressure P_1:

$$F_s = P_1 A_N$$

A slight increase in pressure inside the vessel above P_1 lifts the valve disk up and relieves the vapor through the discharge nozzle of the valve. Accumulation or overpressure above the maximum allowable working pressure (P_1) within the vessel is safe up to about 10 percent, if the overpressure persists for a short period of time. Under normal operation when the operating pressure is much below P_1, the downward force exerted by the spring on the disk is much higher than the upward force exerted by the process vapor. Hence, the disk is held tight against the inlet nozzle under normal operating conditions. When the back pressure exceeds the atmospheric pressure, the combined downward force exerted by the spring and the force developed due of the back pressure is $F_s + P_2 A_N$, where P_2 is the back pressure (refer to Figure 3-1A).

In order to lift the valve disk against this combined downward force, the inlet vapor inside the vessel must be pressurized to a level higher than P_1. Hence, $P_1' A_N = F_s + P_2 A_N$, where P_1' is the new pressure developed.

$$\text{Now max—}P_1' = 1.1\ P_1, \text{ and as noted earlier,}$$
$$U\ F_s = P_1A_N$$

$$\text{Hence, } 1.1\ P_1\ A_N = P_1A_N$$
$$\text{or } P_2 = 0.1\ P_1$$

Thus, the maximum allowable back pressure is 10 percent of the maximum allowable working pressure of the vessel.

The remaining three types of pressure-relief valves do not depend upon the back pressure for their performances. However, to ensure that the safety valves work at their maximum capacity, the back pressure is limited to 50 percent of the relief valve set pressure. In the balanced bellows type valve, the spring does not act directly on the disk. Instead, it serves on a bellows first, which in turn acts on the disk. In case of the piston type, it works on the same principle as the bellows type, excepting that the bellows is replaced by a piston (see Figure 3-1B). The cross-sectional area of both the piston and the bellows is the same as the inlet nozzle of the valve and the effect of the back pressure on the top and the bottom of the disk creates equal balancing forces. That is, P_1A_N is always equal to F_s, as shown in Figure 3-1B.

Pilot-operated valves have a pilot valve combined with the main valve. The spring of the main valve provides 75 percent loading on the disk and the remaining 25 percent is offered by the gas or vapor through the pilot valve. When the vessel reaches the maximum allowable working pressure, the pilot valve relieves the gas pressure, which contributes to the disk load. Thus the safety valve becomes wide open.[2,3] This is illustrated in Figure 3-2.

With all the nonconventional valves, the maximum allowable back pressure may be taken as high as 50 percent of the valve set pressure. This pressure value approaches the critical flow pressure. If the back pressure becomes greater than the critical flow pressure corresponding to the set pressure of the safety valve, the total pressure drop available for flow of the vapors through the safety valve decreases. This state can potentially lead to overpressurizing the vessel. The recommended back pressure is, therefore, a maximum 40 to 50 percent of the set pressure in

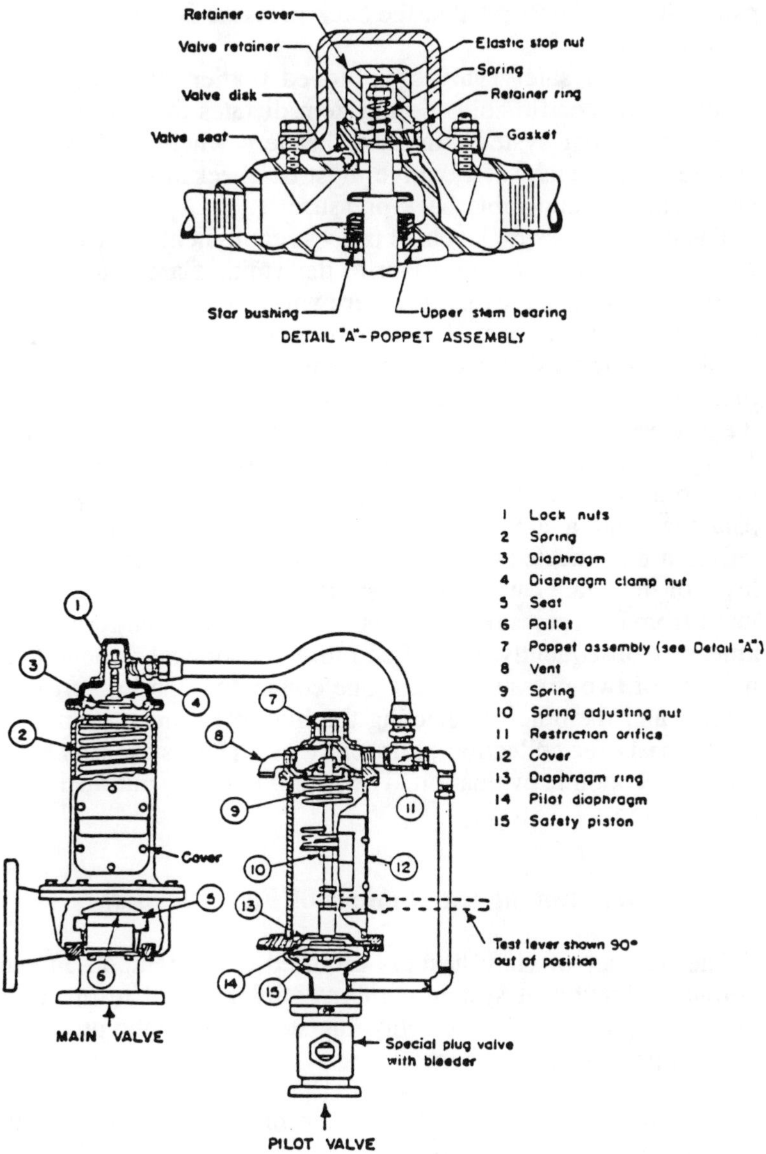

Figure 3-2. Pilot operated safety valve.

psia. This level ensures that the back pressure is below the critical pressure.

The type of safety valves employed (either conventional or others) in a specific collection system dictates the level of back pressure in that system. In flare headers where multiple discharges exist, each safety valve must be checked so that it does not exceed its allowable back pressure.

The design pressure levels of process equipment connected to a flare system are also important to determine flare header pressure levels. In some cases, pressure varies widely from one process vessel to another. Hence, it is not always economical to combine systems with a common header. For example, if the set pressures are 20 psig and 200 psig for two different systems, for the low pressure system the maximum back pressure attainable is 17 psia by using a balanced type safety valve. The high-pressure system is capable of withstanding a back pressure as high as 34 psia even with a conventional type safety valve. But if these two valves are connected to the same header, the maximum pressure level of the header will be 17 psia in order to protect the equipment from overpressure, and this requires a much larger pipe diameter. Consequently, it is often more economical to have flare headers of two pressure levels, one connecting the low-pressure system and the other connecting the high-pressure system.

The basis for collection philosophy of a flare system is based upon an economic evaluation. The methodology is outlined in the next section.

Determining the Number of Flare Headers

The number of flare headers required depends upon an economic evaluation of system combinations that will result in the minimum piping cost. The following steps outline the procedure for comparative estimations:

1. *Study of the Plot Plan Layout:* From the plot plan layout the number of safety valves in different process areas, the set pressures of safety valves, individual relieving loads of safety valves, relieving temperature of vapors, the nature

of vapors (i.e., whether corrosive, condensable, or dry) are recorded.

2. A single subheader in each process area is drawn up, connecting area pressure-relief valves or depressuring valves.
3. The subheaders are then connected to give a single main flare header based upon the shortest routing.
4. The equivalent length of the main flare header is then calculated from the flare stack to the last safety valve, taking into consideration the straight length of the pipe and approximate equivalent lengths for bends, etc. If the actual location of the flare stack is not known by that time, it may be assumed to be 500 ft from the last piece of equipment. Later on, even if it varies from 500 ft, it will not affect the pressure drop calculation at all compared with the entire length of the pipe.
5. A trial estimate is made for determining the diameter of the flare header based upon the maximum relieving flare load and considering the back pressure limitation of 10 percent for conventional valves and 40 percent for balanced type valves. Note, however, a single main header in most cases turns out to be too large to be economically feasible. Line sizing procedures are discussed in detail in the next section.
6. The second trial is required for two main flare headers, one collecting the low-pressure flares (usually 5 to 10 psig) and the other collecting relatively high-pressure flares (usually 15 to 20 psig). The two headers are connected to their individual knock-out drums. The vapor lines from the knock-out drums are combined into a single header connected to the flare stack.

   The maximum simultaneous load in each header must be calculated separately and the pressure drop must also be computed for the entire length of the pipe including the combined length from the knock-out drum to the stack.

   The subheaders in each process area similarly will have two levels of flare headers. The line sizing of each level of subheader in an individual area will depend upon the maximum simultaneous flow in that particular area. Thus the line sizing criterion of a subheader may be the largest single flow due to a blocked outlet condition. This flow may not necessarily be the controlling load for the flare stack.

7. The next consideration is the cost of construction materials. This determines the final number of flare headers. Vapors that normally require expensive materials may be listed as follows:

a. Corrosive vapors, e.g., SO_2, H_2S.

b. Very high temperature vapors, e.g., high temperature gases used for regeneration of catalysts in reactors.

c. Very low temperature vapors, e.g., vapors generated due to flashing across a control valve or a safety valve in a cryogenic system.

Of the three, corrosive vapors are usually piped up in a separate header right up to the flare stack since such lines are normally very small and if combined with other streams may run the risk of corroding the much larger and more expensive pipelines.

For a high-temperature system, a separate subheader may be run up to the point where the temperature drops down to the allowable limit of a less expensive material. It may then be connected to the main flare header (either low pressure or high pressure). A heat loss calculation is needed in order to properly evaluate this. As a rule of thumb a heat loss of 10 BTU/hr/ft^2 may be assumed for a quick estimate for bare pipe. Consideration should also be given to the need for expansion joints. Main flare headers may be as large as 36 to 42 inches in diameter for a large-capacity plant. Expansion joints of such magnitudes may be so expensive as to call for a separate small header for the hot flare system.

The flare subheaders carrying very low temperature vapors (temperatures ranging from 50°F and below) may similarly be combined into a single low temperature flare header and piped all the way up to the flare stack. Again, since the atmosphere is warmer than the pipes, a heat gain calculation may indicate that the cold subheaders after running a certain distance by themselves may be safely combined either with the low-pressure main flare header or the high-pressure main flare header, depending upon their operating pressure.

8. *Wet Flare and Dry Flare:* Sometimes relatively hot vapors carrying condensates may be separated from the dry cold vapors. They do not run as separate headers but either low-pressure or high-pressure flare headers may be associated with any one of them. Thus a wet flare header may be, in fact, the low-pressure header, and the dry flare header may be the high-pressure flare, or vice-versa.
9. After the total number of flare headers has been established, it may be necessary to recheck the vapor load in individual headers since introduction of a separate header may allow subtraction of the flow quantity from the low-pressure header to which it was added initially.

Two examples are an ethylene plant and a coal gasification plant.

A typical ethylene plant usually has the following flare headers:

a. Low-pressure wet flare header containing hot gas and water.

b. High-pressure dry flare header containing cold gas and devoid of water.

c. Liquid drain header containing low-temperature vapors after flashing across valves.

A typical coal gasification plant has the following flare headers:

a. High-pressure dry flare header.

b. High-pressure wet flare header.

c. An H_2S header containing vapor of more than 5 percent (volume of H_2S).

Line Sizing for Flare Headers

When the maximum vapor-relieving requirement of the flare system has been established and the maximum allowable back pressure (as just described) has been defined, line sizing reduces to standard flow calculations.

The flare piping system can be divided into the following sections:

1. Individual discharge lines from the pressure-relief valves.
2. The subheaders in each area connecting the discharge lines.
3. The main flare header connecting the subheaders leading to the knock-out drum.
4. The final header connecting the vapor line (or lines) from the knock-out drum leading to the flare stack.

Since vapors in the flare headers are relieved from a high pressure system to almost atmospheric pressure, there is an appreciable kinetic energy change throughout the pipeline. The flow condition is that of compressible flow. The nature of compressible flow in the case of flare headers may be assumed to be isothermal since flare lines are normally long and not fully insulated. Short and properly insulated vapor lines are close to an adiabatic flow. In general, all vapor flows that normally occur in process plants are somewhere in between adiabatic and isothermal flow. It has been observed that for the same flow rate and pressure drop, line sizing calculations based on compressible isothermal flow conditions always give an equal or a larger diameter pipe. Hence, flare headers should be sized based upon isothermal compressible flow for a more conservative design.

The following criteria are used in sizing flare headers:

1. The back pressure developed at the downstream section of any pressure-relief valve connected to the same headers should not exceed the allowable limit, i.e., 10 percent of the set pressure in psig for the conventional type and 40 to 50 percent of the set pressure in psia for the balanced type valve.

2. Since the pressure drop is quite high, there is a possibility of approaching the sonic velocity in the line. This will result in a potential noise problem. Hence, it is a good practice to limit the velocity to 60 percent of the sonic velocity or 0.6 Mach number.

A quick method for sizing compressible isothermal flow is offered by the following method developed by Lapple.[4]

This method employs a theoretical critical mass flow based on an ideal nozzle and isothermal flow condition. For a pure gas, the mass flow can be determined from one equation:

$$G_{ci} = 12.6\,P_o\left(\frac{M}{(2Z - 1)T_o}\right)^{0.5} \qquad (3\text{-}1)$$

where

G_{ci} = maximum mass flow or critical mass flow, in lbs per sec per sq ft
P_o = upstream pressure, in lbs per sq in. absolute
M = mol wt
T_o = upstream temp in °R
Z = compressibility factor

The actual mass flow through a pipe, G, in lbs per sec per sq ft is a function of critical mass flow G_{ci}, line resistance, N, and the ratio of downstream to upstream pressure. These relationships are plotted in Figure 3-3. In the area below the dashed line in Figure 3-3, the ratio of G to G_{ci} remains constant, which indicates that sonic flow has been established. Thus, in sizing flare headers the plotted point must be above the dashed line. The line resistance, N, is given by the equation:

$$N = \frac{4fL}{D} + \Sigma K_i \qquad (3\text{-}2)$$

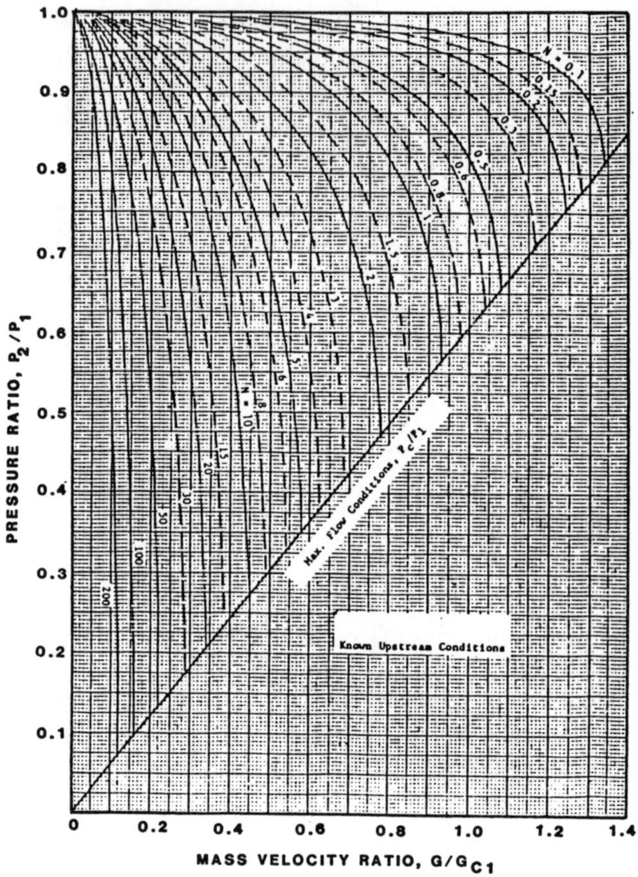

Figure 3-3. Pressure drop chart—after Lapple.[4]

where

L = equivalent length of line, in ft
D = line diameter, in ft
f = Fanning friction factor
N = line resistance factor, dimensionless
K_i = resistance coefficients for pipe fittings (see Table 3.1).

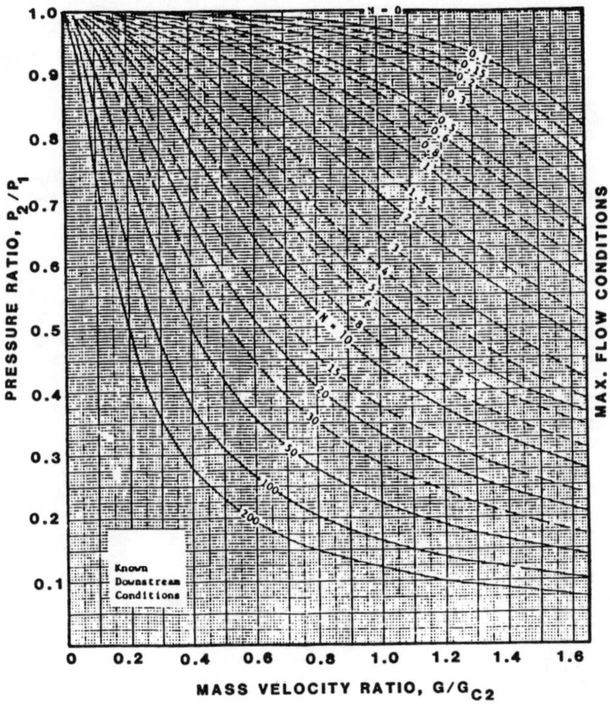

Figure 3-4. Pressure drop chart—after Loeb.[5]

Lapple's method is useful when the upstream pressure of a header is known and the downstream pressure has to be calculated. However, it is often required to develop the pressure profile of the flare headers as a function of the distance from the stack. For this reason, it is more convenient to calculate the pressure drop backward, starting from the flare stack exit where the pressure is atmospheric. Figure 3-4 provides another plot which enables the pressure loss calculation when the downstream pressure is known.[5]

Table 3-1
Resistance Coefficient K For Various Pipe Fittings

Fitting	*K*	Fitting	*K*
Globe valve, open	9.7	90° double-miter elbow	0.59
Typical depressuring valve, open	8.5	Screwed tee through run	0.50
Angle valve, open	4.6	Fabricated tee through run	0.50
Swing check valve, open	2.3	Lateral through run	0.50
180° close screwed return	1.95	90° triple-miter elbow	0.46
Screwed or fabricated tee through branch	1.72	45° single-miter elbow	0.46
90° single-miter elbow	1.72	180° welding return	0.43
Welding tee through branch	1.37	45° screwed elbow	0.43
90° standard screwed elbow	0.93	Welding tee through run	0.38
60° single-miter elbow	0.93	90° welding elbow	0.32
45° lateral through branch	0.76	45° welding elbow	0.21
90° long-sweep elbow	0.59	Gate valve, open	0.21

d/d′:	*0*	*0.2*	*0.4*	*0.6*	*0.8*
Contractions (USASI)	—	—	0.21	0.135	0.039
Contractions (sudden)	0.5	0.46	0.38	0.29	0.12
Enlargements (USASI)	—	—	0.9	0.5	0.11
Enlargements (sudden)	0.0	0.95	0.74	0.41	0.11

Although Figures 3-3 and 3-4 can be used for line sizing, it should be noted that Figure 3-3 requires more extensive trial and error calculations.

The following steps summarize the procedure for sizing flare headers:

1. The pressure at the base of the flare stack is approximated as 2 psig. Depending on the type of seal used in the flare stack, the pressure at the base may vary slightly although 2 psig is generally a good approximation:

Pressure at the base = Atm pressure at the flare exit + 0.5 psi flare tip

ΔP + 0.5 psi mol seal ΔP + 1 psi
ΔP due to flow through the stack height ≃ 2 psig

2. Compute the pressure in the knock-out drum:

= 2 psig + ΔP required for the flow of the full load vapors from the knock-out drum to the stack + 0.5 psi ΔP assumed inside the K.O. drum.

In the absence of actual line distances, a conservative estimate of the distance between the knock-out drum and the flare may be taken as 500 ft.

3. As a first trial, an inside pipe diameter is assumed based on 60 percent of the sonic velocity corresponding to the pressure and temperature at the base of the stack, i.e., at 2 psig and temperature = T_o (upstream temperature since isothermality is assumed).

The sonic velocity can be computed from:

$$V_s = 223\sqrt{\frac{KT}{M}} \tag{3-3}$$

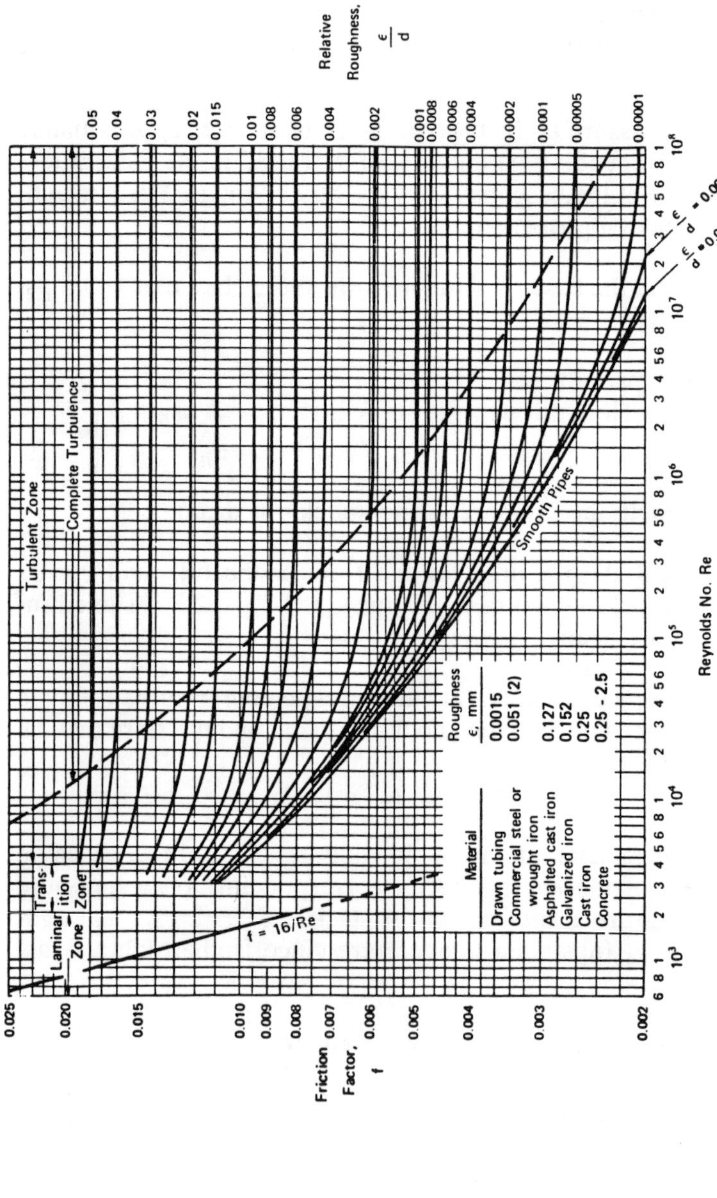

Figure 3-5. Generalized Moody plot for obtaining friction factor.[6]

where

V_s = sonic velocity in ft/sec
K = C_p/C_v of the gas; normally between 1 to 1.8
T = temperature in °R
M = mol. wt.

The flare load, W, lbs/sec. is known. The density of the vapor at 2 psig, T_o °R is determined from the ideal gas law:

$$P = \frac{MP}{RT}$$

where

M = mol. wt.
P = pressure in psia
T = temperature, °R
R = gas constant

Hence,

$$\frac{W}{\pi d^2/4} = 0.6 \times 223 \sqrt{\frac{KT}{M}} \quad (3\text{-}4)$$

where

d = pipe ID in ft

The inside diameter d can be calculated from Equation 3-4 since everything else is known.

Once the diameter is known, the Reynolds number Re can be computed and the friction factor f obtained from Figure 3-5. Assuming a straight length of pipe for L = 500 ft, N (line resistance factor) can be calculated. Next G_{ci} is calculated based on the downstream pressure and G/G_{c2} evaluated. From Figure 3-4 the ratio P_2/P_o can be obtained. Since P_2 is known, P_o can then be calculated. The pressure at the inlet of the knock-out drum is

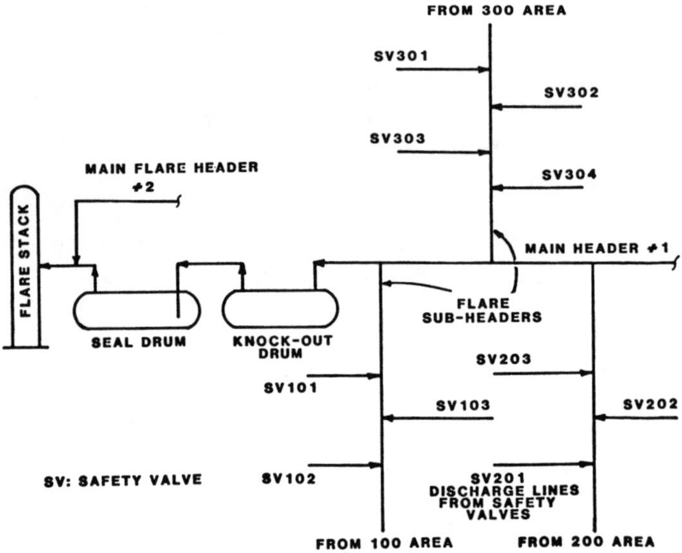

Figure 3-6. Typical layout of a flare collection system.

given by $P_o + 0.5$ psi. Table 3-1 provides typical values of resistance coefficients for various pipe fittings. Details of pressure loss calculations are given in Chapter 6.

4. From the knock-out drum, the individual flare headers can now be sized in the same manner. Based on a Mach number of 0.6 and the density corresponding to $(P_o + 0.5)$ psia, a trial diameter can be established. The pressure at every intersection between the subheader and the main header must be calculated, with the downstream pressure being $(P_o + 0.5)$ psia. Knowing the pressure at the intersection of the subheader and the main header, the pressure at the intersection of the subheader and the discharge pipe of the safety valve is computed. Usually the discharge pipe of the safety valve is small and hence, a separate

pressure drop calculation is not necessary. However, the velocity at the discharge pipe should be checked to ensure that it is below sonic velocity. If the discharge pipe runs a considerable distance before it ties in with the subheader, a separate pressure drop calculation will be required. Figure 3-6 shows a typical layout of the flare piping.

The sum of all the pressure losses starting from the flare stack up to the safety valve yields the total back pressure in psig. This back pressure must be lower than the maximum back pressure allowed in the system and corresponding to the lowest set pressure of the safety valve.

The following examples illustrate the calculations needed to properly size lines for flare headers.

Example 3-1: The maximum flare load of a system is 1,000,000 lbs/hr of vapor. The pressure at the base of the flare stack is 2 psig, the average M.W. of the vapor is 50, temperature 200°F at the combined header to the flare stack. The distance from the drum to the stack is 500 ft. The line consists of two 90° welding elbows and an orifice for a flow controller. The total pressure drop at the knock-out drum is 0.5 psi. Determine the pressure at the inlet of the knock-out drum.

Additional Data:

$K = C_p/C_v = 1.2$
Z (compressibility factor) $\simeq 1$

Trial No. 1 (From Equation 3-4):

$$\frac{W}{\rho \pi d^2/4} = 0.6 \times 223 \sqrt{\frac{KT}{M}}$$

$$\frac{1{,}000{,}000 \times 4}{3600 \times 0.125 \times 3.14d^2} = 0.6 \times 223 \sqrt{\frac{1.2 \times 660}{50}}$$

$$\rho = \frac{MP}{10.73T} = 0.125 \text{ lbs/cu ft}$$

$$d = 2.3 \text{ ft} = 27.7 \text{ in.} \approx 29 \text{ in.}$$ corresponding to standard pipe 30 in., 20 schedule

Compute Line Resistance Factor (N):

Since Fanning's friction factor "f" does not vary much, f = 0.004 is a good approximation. However, it can be calculated from Re by the standard method from Table 3-2.

$$\Sigma K = \underset{\text{Elbow}}{0.32 \times 2} + \underset{\text{Sudden Enlargement}}{0.2}$$

$$= 0.84$$

$$N = \frac{4fL}{D} + \Sigma K_i$$

$$= \frac{4 \times 0.004 \times 500 \times 12}{29} + 0.84$$

$$= 4.04$$

Determine G/G_{ci}:

$$G = \frac{1{,}000{,}000 \times 4 \times 12 \times 12}{3600 \times 3.14 \times 29 \times 29} = 56.6 \text{ lbs/sec/ft}^2$$

$$G_{ci} = 12.6\, P_o \left[\frac{M}{(2Z - T_o)}\right]^{0.5} \text{ (From Equation 3-1)}$$

The upstream pressure P_o will be replaced by the downstream pressure = 2 + 14.7 = 16.7

$$Z = 1,\ T_o = 200 + 460$$

$$= 660°R \text{ (isothermal)}$$

$$\text{Hence, } G_{ci} = G_{c2} = 12.6 \times 16.7 \sqrt{\frac{50}{660}}$$

$$= 57.97 \text{ lbs/sec/ft}^2$$

(G_{c2} represents mass flow rate based on downstream conditions)

$$G/G_i = G/G_{c2} = \frac{56.6}{57.92} = 0.977$$

$$P_2/P_1 = 0.59 \text{ (From Figure 3-4)}$$

$$P_1 = \frac{16.7}{0.59} = 28.3 \text{ psi}$$

$\Delta P = P_1 - P_2 = 28.3 - 16.7 = 11.6$ psi which is too high (as a rule of thumb, ΔP should not exceed 3 psi). This means that a larger line size should be chosen.

Trial No. 2

Assume an inside diameter of 35.25 in. (i.e., nominal 36 in. pipe, standard thickness).

$$N = 3.5,\ G/G_{c2} = 0.678$$

$$P_2/P_1 = 0.79,\ P_1 = 21 \text{ psi},\ \Delta P = 4.3 \text{ psi}$$

Trial No. 3

$$ID = 39.25 \text{ in., (40 in. std.)}$$

$$N = 3.24,\ G/G_{c2} = 0.55$$

$$P_2/P_1 = 0.85,\ P_1 = 19.6 \text{ psi},\ \Delta P = 3.3 \text{ psi}$$

Trial No. 4

ID = 41.25 in., N = 3.125, G/G_{ci} = 0.483 (Corresponding to 42 in., standard thickness)
P_2/P_1 = 0.87, P_1 = 19.2
ΔP = 2.5 psi

Hence, total pressure drop

= 2.5 + 0.5 + 0.25
= Line ΔP + K.O. drum + orifice ΔP

ΔP = 3.25 psi

The pressure at the inlet of the knock-out drum is 16.7 + 3.25 =

19.95 psi or
20 psi

Example 3-2

Calculate the dry flare header size, which is connected to the above knock-out drum.

Previously noted conditions remain the same. Additional data are:

W = 720,000 lbs/hr, Temperature = 100°F

and the maximum allowable back pressure at the safety valve (as shown in Figure 3-7) is 34 psia. The discharge rate from the safety valve is 30,400 lbs/hr.

Trial No. 1

Assuming a Mach number of 0.6 and d ≃ 21.25 in. (22 in. standard).

$$\rho = \frac{50 \times 20}{10.73 \times 560} = 0.169 \text{ lbs/cu. ft}$$

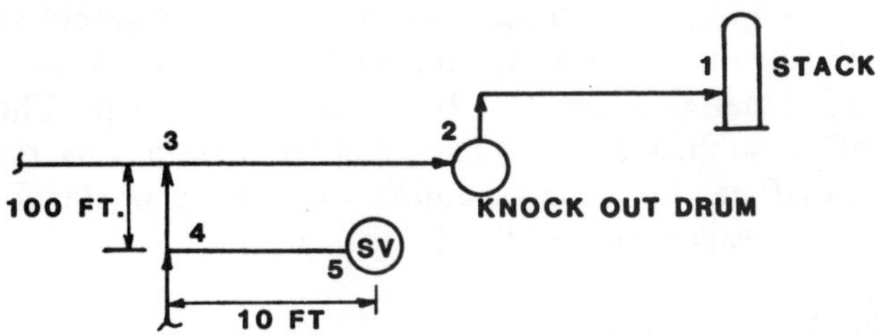

Figure 3-7. System definition for Example 3-2.

$G/G_{ci} = 0.988$, $N = 4fL/D = 17.45$, P_2/P_1 from Figure 3-4 = 0.36, $P_1 = 56.5$ psi, which is too high.

Trial No. 2

Use ID = 29 in.
$G/G_{ci} = 0.53$, $N = 12.8$, $P_2/P_1 = 0.66$
$P_1 = 30.8$ psi, $\Delta P = 30.8 - 20$ ft
= 10.8 psi

Trial No. 3:

ID 27.25 in. (28 in. Standard)
$G/G_{ci} = 0.568$, $N = 13.7$, $P_2/P_1 = 0.61$
$P_1 = 33.36$ psi

Hence, either a 28 or 30 in. diameter is acceptable for the main header size depending on the ΔP at the subheader.

We must now consider the pressure drop between points 4 and 5 in Figure 3-7. Since the distance between points 4 and 5 is only 10 ft, ΔP calculations are not warranted. Instead, we can assume the same diameter for the subheader and the discharge line of the safety valve. It is quite possible that because of some other safety valves discharging to the same subheader, the load on the subheader is much higher than the given rate from the safety valve, i.e., 30,400 lbs/hr. As a result, the subheader size should actually be larger than that specified based on 30,800 lbs/hr. Thus, if the size of the subheader is not known at the time, a conservative estimate of ΔP can be made assuming a smaller diameter for the subheader corresponding to 30,400 lbs/hr.

Trial No. 1

$$\rho = \frac{50 \times 33.36}{10.73 \times 560} = 0.277 \text{ lbs/cu ft @ } 33.36 \text{ psia and } 100°\text{F}$$

corresponding to 28 in. ID for main header

For a Mach number of 0.6, d = 3.3 in.

But the outlet nozzle size of the safety valve is 6 in. Schedule 40. Hence, 6 in. will be minimum line size. (This is normal practice.)

ID = 6.065 in.

$G/G_{ci} = 0.34$, $N = 3.52$, $P_2/P_1 = 0.94$, and
$P_1 = 35.4$ psi

Trial No. 2:

ID = 7.981 in. corresponding to 10 in., Schedule 40 pipe
$G/G_{ci} = 0.192$, $N = 2.65$, $P_2/P_1 = 0.98$
$P_1 = 34.1$ psia

In summary, we have obtained the following:

Combination No. 1—42 in. of 500 ft, 28 in. of 2,000 ft, 8 in. of 110 ft, pressure at point 5 = 34.1 psia

Combination No. 2—40 in. of 500 ft, 30 in. of 2,000 ft, 8 in. of 110 ft, pressure at point 5 = 32.15 psia

Combination No. 3—40 in. of 500 ft, 32 in. of 2,000 ft, 6 in. of 110 ft, pressure at point 5 = 32.82 psia

Combination No. 1 appears to be the most economical since the 28 in. line has the maximum run. Next is No. 2, compared to No. 3 for the same reason.

4

DESIGNING THE FLARE STACK AND ACCESSORIES

Knock-Out Drum Sizing

The hydrocarbon relief streams are mainly vapors, but they may carry some liquids that condense in the collecting line. Therefore, material entering the knock-out drum (also called a blow-down drum) will be a mixture of vapor and liquid. A particle that is 150 microns or less can be burned in the flare without hazard. Larger particles must be removed in the knock-out drum. This liquid is pumped out from the bottom of the knock-out drum either for reuse or for disposal in a slop tank. Figure 4-1 illustrates a typical knock-out drum with pump-out connections.

In some process plants, e.g., ethylene plants or coal gasification plants, hot vapors containing water are collected in a separate flare header (called a wet flare header). The liquid collected in the knock-out drum for the wet flare contains water and liquid hydrocarbons. In the same manner, cold and dry hydrocarbon vapors are collected in a dry flare header. The hydrocarbon liquid collected in the knock-out drum of the dry flare is usually vaporized below the knock-out drum and sent back to the flare. Figure 4-2A shows a dry flare knock-out drum with a vaporizer at the bottom.

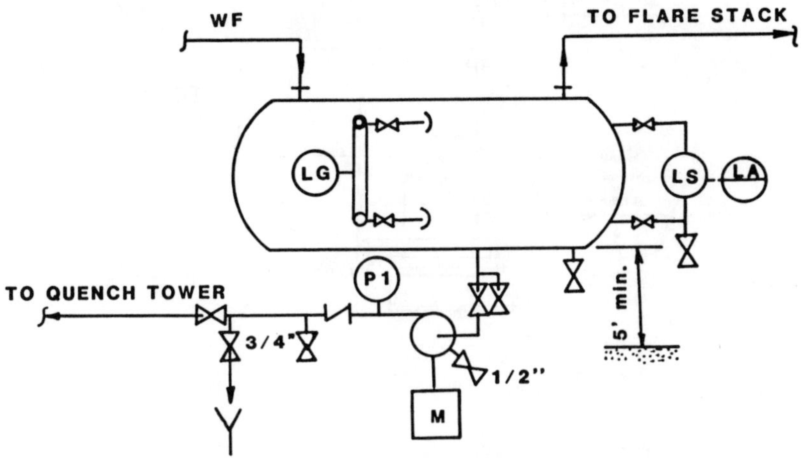

Figure 4-1. Typical knock-out drum with level gauge, level switch and alarm. Also shows a 2-in. utility connection for cleaning the drum with steam and a pump-out pump with a pressure indicator and a drain connection.

Knock-out drums are either horizontal or vertical. They are also available in a variety of configurations and arrangements that include:

1. A horizontal drum with the vapor entering at one end of the vessel and exiting at the top of the opposite end (no internal baffling).
2. A horizontal drum with the vapor entering at each end on the horizontal axis and a center outlet.
3. A horizontal drum with the vapor entering in the center and exiting at the two ends on the horizontal axis.
4. A vertical drum with the vapor entering at the top on a certain diameter and provided with a baffle so the flow is directed downward. The outlet nozzle is located at the top of the vertical axis (see Figure 4-2C).
5. A vertical drum with a tangential nozzle.

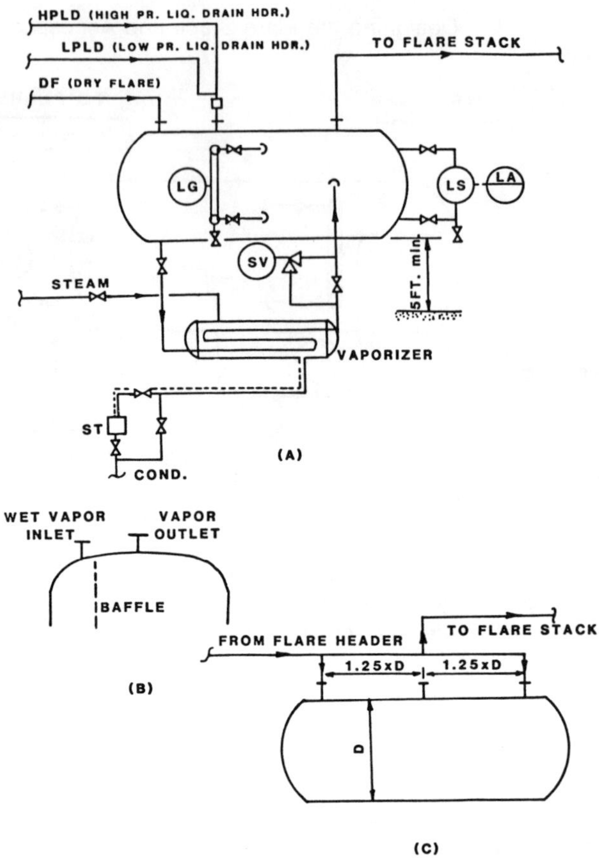

Figure 4-2. (A) Dry-flare knock-out drum. (B) Alternate split feed arrangement of knock-out drum. (C) Shows vertical drum arrangement.

Selection of the drum arrangement depends on economics. When large liquid volume storage is required and the vapor flow is high, normally a horizontal drum is more economical.

A split entry or exit reduces the size of the drum for large flows. As a rule of thumb, when the drum diameter exceeds 12 ft, the split flow arrangement is normally economical. Figure 4-2B shows a split flow horizontal drum with the recommended dimensions.

Knock-out drums are usually sized by a trial and error method. Liquid particles drop-out when the vapor velocity traveling through the drum is sufficiently low. In other words, the drum must be of sufficient diameter to effect the desired liquid-vapor separation.

Tan[8] gives the following formula for sizing horizontal drums:

$$W = 360D^2 \sqrt{(\rho_L - \rho_G)\, MP/T} \tag{4-1}$$

Valid for particle size of 400 microns

where

W = lbs/hr of vapor
ρ_L = liquid density lbs/cu ft
ρ_G = gas density lbs/cu ft
M = mol wt of the vapor
T = temperature of the vapor in °R
P = psia, D in ft
D = drum diameter, ft

Similar expressions are available for vertical knock-out drums. A practical formula for the vapor velocity is:

$$V = 0.4 \sqrt{\frac{\rho_L - \rho_G}{\rho_G}}, \quad \text{ft/sec} \tag{4-2}$$

The following example illustrates the calculations for drum sizing:

Example 4-1

The flow rate of a vapor stream is:

290,000 lbs/hr Temp = 500°F summer temperature
Pressure = 3.5 psig,
Mol. Wt. = 50, ρ_L = 40 lbs/cu. ft, ρ_G = 0.08 lbs/cu. ft

A liquid holdup of 20 minutes is specified, which is equivalent to 500 gallons.

First Trial: Assume a horizontal drum and similar to type 1 which is the simplest type.

$$\text{Now, } W = 360D^2 \sqrt{(\rho_L - \rho_G)\ MP/T}$$

$$290{,}000 = 360D^2 \sqrt{(40 - 0.08)\ \frac{50 \times 18.2}{960}}$$

$$D = 11.44 \text{ ft}$$

To convert to the diameter corresponding to a 150μm particle size, the following relationship may be used:

$$150 = 400 \left(\frac{D_o}{D}\right)^4$$

where D_o – dia corresponding to 400μ

hence

$$D = 14.6 \text{ ft, vapor velocity} = 6 \text{ ft/sec.}$$

This diameter is for an empty vessel, (i.e., no liquid). In order to maintain a liquid level, the diameter must be larger. The diameter for the simplest drum type will be too large. Hence, a split flow arrangement is necessary.

Since the flow of vapor will be divided, the new cross-sectional area with the same vapor velocity will be equal to:

$$\frac{29{,}000}{2 \times 0.08 \times 3600 \times 6} = 83.91 \text{ sq ft}$$

Hence, diameter = 10.33 ft, which corresponds to an empty vessel.

It is a general practice to assume a liquid holdup time between 10 and 30 minutes. In the absence of data, a volume of 2,000 gals of liquid will be a good approximation (i.e., 267.38 ft^3 of liquid).

Assume a tank diameter of 11 ft, with a length of 30 ft.

Cross section occupied by the liquid $= \dfrac{267.38}{30} = 8.91$ sq ft.

Cross-sectional area of the drum $= \pi/4 \times 121$ sq ft.

Cross-sectional area occupied by vapors $= 94.98 - 8.91 = 86.07$ sq ft

This is greater than the required area of 83.91 sq ft, just calculated.

The minimum L/D ratio recommended for a split flow horizontal drum is 2.5 (see Figure 4-2B) for proper separation of liquid particles from vapors. The ratio used in this drum size $= \dfrac{30}{11} = 2.727$.

Hence, the dimensions of the tank as assumed (11 ft dia × 30 ft length) are acceptable.

For Vertical Flow:

$$\text{Velocity} = 0.4\sqrt{\frac{\rho_L - \rho_G}{\rho_G}}\ \text{ft/sec}$$

$$= 0.4\sqrt{\frac{40 - 0.08}{0.08}}\ \text{ft/sec}$$

$$\text{Hence, diameter of the drum} = \frac{290{,}000 \times 4}{3600 \times 0.08 \times 3.14 \times (8.9)^2}$$

$$= 16.19\ \text{ft}$$

The diameter is too large and will not be economical compared to the horizontal split flow type.

To determine the liquid level in the drum:

A_S = cross-sectional area occupied by the liquid = 8.91 sq ft
R^2 = diameter square of the drum = 30.25 sq ft

$$\frac{A_S}{R^2} = 0.3 \text{ (see Perry's } \textit{Chemical Engineering Handbook.}^9\text{)}$$

H/R corresponding to $\frac{A_S}{R^2} = 0.3$

Hence, liquid level $H = 0.3 \times 5.5 \times 12 \simeq 20$ in.

Seal System

Standard practice in refineries and most chemical plants is to provide a seal at the base of the flare to prevent flashbacks. In the absence of a seal, a continuous quantity of gas may be bled to the flare to maintain a positive flow. Seals are of two main types: liquid seal and gas seal. Both are widely used.

Liquid seals are further classified as seal drums and seal pipes. In the former, a liquid seal is used in a seal drum located between the knock-out drum and the flare stack. Instead of a drum, sometimes a piping seal is used as a seal leg located at the bottom of the flare stack. This is often an integral part of the stack. Seal drums are also either vertical or horizontal (see Figure 4-3A and 4-3B). The selection of the seal drum depends essentially on the availability of space. The purpose of a seal drum is to maintain a seal of several inches on the inlet flare header, preferably not exceeding six inches—otherwise it causes a back pressure on the knock-out drum. Water is normally used as a sealing liquid and there is always a continuous flow of water with the overflow going to the sewer. If located in a cold climate, the water must either be heated by a submerged steam heater or it may be replaced by liquids such as alcohol, kerosene, etc., which do not require continuous flow.

The capacity of the seal drum is usually the volume corresponding to 8 to 10 ft of the vapor inlet line. In a vertical drum, the ratio of the inlet pipe cross-sectional area to the vessel-free area for gas flow above the liquid should be at least 1 to 3 to prevent

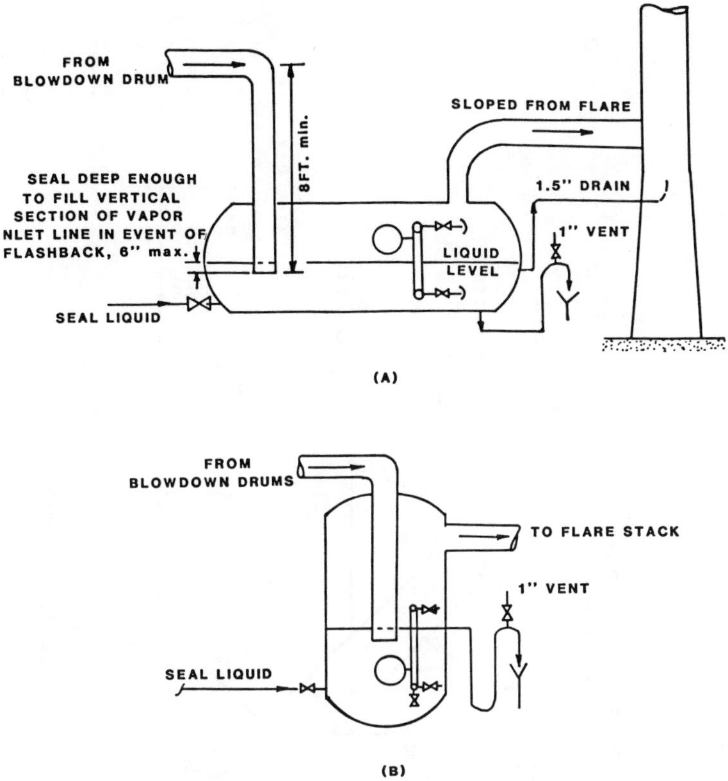

Figure 4-3. (A) Shows a horizontal seal drum. (B) A vertical seal drum.

upsetting surges of gas flow to the flare. The area for the gas above the liquid surface should be at least equal to that of a circle having a diameter D equal to 2d, where d = inlet gas pipe diameter. This can be derived as follows:

Assuming a vertical vessel of cross-sectional area $\frac{\pi}{4} D^2$ and the inlet pipe $\frac{\pi}{4} d^2$, the annular area is thus $\pi/4\ (D^2 - d^2)$. Since the suggested ratio is 1:3, $D^2 - d^2 = 3d^2$ or $D = 2d$.

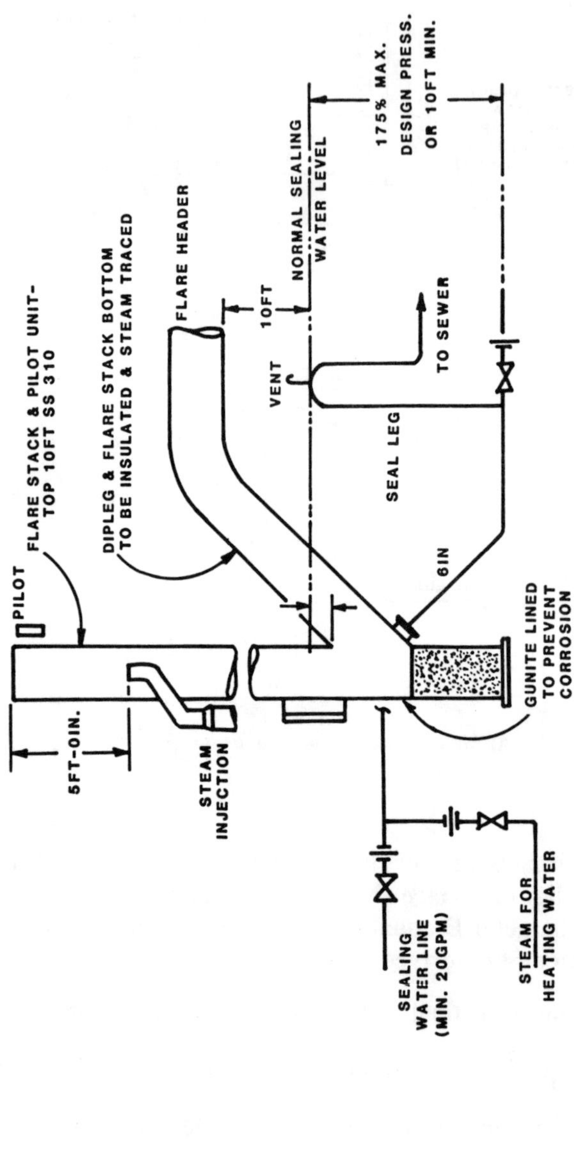

Figure 4-4. Seal leg arrangement.[9]

The height of the vapor space above the liquid level in a vertical drum should be approximately two to three times the diameter (d) to provide disengaging space for entrained seal liquid. Figure 4-5 shows a typical vertical seal drum.

If a horizontal seal vessel is used, a minimum dimension of 3 ft between liquid level and top of the drum is recommended.

Seal pipes located at the base of the stack are cheaper than drums. However, they can experience pulsation of the gas flow to the flare under very low flow conditions. Also, during a large gas release, the water seal may be blown out of the top to the flare stack. Some general guidelines for sizing seal legs follow:

1. Slope of the inlet line is designed to provide a volume of water below the normal sealing water level equivalent to the inlet pipe volume of 10 ft.
2. Depth of the water seal should not exceed 12 in. to prevent gas pulsation.
3. Seal water level is maintained by a continuous flow of water at about 20 gallons per minute.
4. Normal overflow is taken off the bottom of the seal through a seal leg. The height of the seal leg should be equivalent to about 175 percent of the pressure at the base of the stack during maximum vapor release so that gas release at the base of the flare is prevented. Figure 4-4 shows a simplified seal leg arrangement.

A more recent *gas seal* type of device that has been developed to prevent flashbacks and explosions in the flare system is the "molecular" type seal. It uses a purge gas of molecular weight 28 or less (e.g., N_2, CH_4, or natural gas). Because of the buoyancy of the purge gas, it creates a zone having a pressure greater than the atmospheric pressure. The molecular seal is located at the top of the flare stack immediately before the burner tip. The ambient air cannot enter the stack because of this high pressure. The recommended purge velocity through the molecular seal is about 0.1 ft/sec. If a molecular seal is not used, the recommended velocity is 1 ft/sec, thereby increasing the purge gas requirement (Figure 4-5).

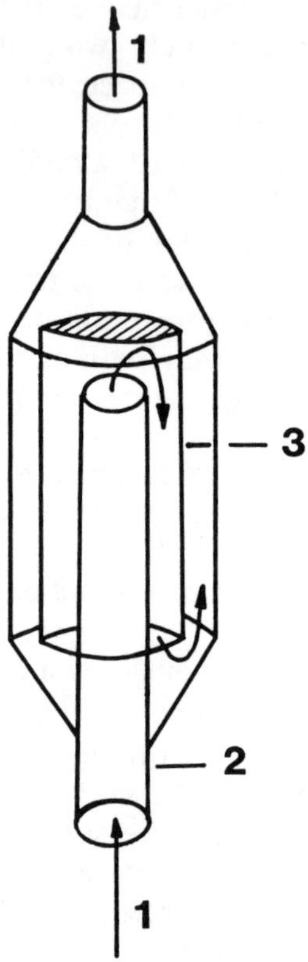

Figure 4-5. Shows a molecular seal.

Flare Burners

The flare burner is located at the tip of the flare stack. The top section is about 12 ft long and is called the flare burner tip. The burner diameter is sized on a velocity basis. Experiments have shown that flame blowout occurs when the exit velocity of vapors exceeds 20 to 30 percent of the sonic velocity.[8] It is therefore good practice to size the burner or the flare stack on a basis of 20 percent of the sonic velocity as the exit velocity. The equation of a burner diameter can then be derived as follows.

The mass flow is given by:

$$W = 3600\ \rho_G\ A_c V \qquad (4\text{-}3)$$

where

W = mass flow rate, lb/hr
ρ_G = density of the gas, lb/cu. ft
V = exit velocity, ft/sec
A_c = cross-sectional area, sq ft

The vapor density is:

$$\rho_G = \frac{MP}{10.73T}$$

The exit velocity equals 20 percent of sonic:

$$V = \frac{1}{5}\sqrt{\frac{gKRT}{M}} \qquad (4\text{-}4)$$

The flare tip cross section is:

$$A_c = \frac{0.785}{144}d^2$$

where

M = mol. wt.
P = absolute pressure of vapor 14.7 psia
T = temperature, °F

g = acceleration of gravity, 32.17 $\dfrac{\text{(lb mass) (ft)}}{\text{(lb force sec}^2\text{)}}$

R = gas constant, 1,546 $\dfrac{\text{(ft lb force)}}{\text{(°R) (mol)}}$

$K = C_p/C_v = 1.2$ assumed
d = diameter of flare tip, in.

Combining these equations and using the values for g, K, R, and P, we obtain:

$$d^2 = \frac{W}{1370}\sqrt{\frac{T}{M}} \tag{4-5}$$

If based upon the maximum rate, the diameter will be too large, and then the normal flow must be used and checked for 40 percent of the sonic velocity for the maximum load. The following sample calculation illustrates the important principles.

Example 4-2

Using the flare load as given in Table 2-1, (i.e., 1,000,000 lbs/hr flowing, temp. = 300°F and mol. wt. = 50), the diameter of the burner (or the stack) becomes 53.3 in. This is too large. It should be sized based on the normal load and checked for the maximum velocity of 40 percent of sonic based upon the maximum load.

$$\begin{aligned} \text{Normal load} &= 800{,}000 \text{ lbs/hr} \\ d &= 47.7 \text{ in.} \\ &\simeq 48 \text{ in.} \end{aligned}$$

Based upon the maximum flow:

$$\begin{aligned} \text{Velocity} &= \frac{1{,}000{,}000 \times 4}{3600\ \rho_G\ 3.14 \times (4)^2} \\ &= 245 \text{ ft/sec} \\ \rho_G &= \frac{MP}{10.73T} \\ &= \frac{50 \times 14.7}{10.73 \times 760} \\ &= 0.09 \text{ lbs/cu ft} \end{aligned}$$

From Equation 4-4:

$$V_s = \sqrt{\frac{gKRT}{M}}$$

$$= \sqrt{\frac{32.2 \times 1.2 \times 1546 \times 760}{50}}$$

$$= 952.8 \text{ ft/sec}$$

Maximum velocity is 25.7 percent of sonic.

Hence, the selected diameter of the burner is 48 in. and it is the same as the stack diameter.

Flare Stacks

The location of the flare stack is a safety-related issue. Normally, it is located in areas on the lee side of the plant (downwind of prevailing winds) and remote from operating and trafficked zones.

The height of the flare stack depends upon the following:

1. Heat released by the flare gas in BTU/hr.
2. Characteristics of the flame and the flame length.
3. Emissivity of the flame.
4. Radiation intensity of the flame in BTU/hr/ft^2.
5. Ground level concentration of toxic gases present in the flare stream in the event of a flame blowout.

There is considerable interest in flame mechanics. Flame burning characteristics and flame length are of considerable importance in sizing the flare stack.

Flame burning characteristics are shown in Figure 4-6, which identifies zones of the flame spectrum in terms of dimensionless numbers. Figure 4-7 enables estimations of the critical flame points in each combustion zone. Figure 4-8 helps to visualize

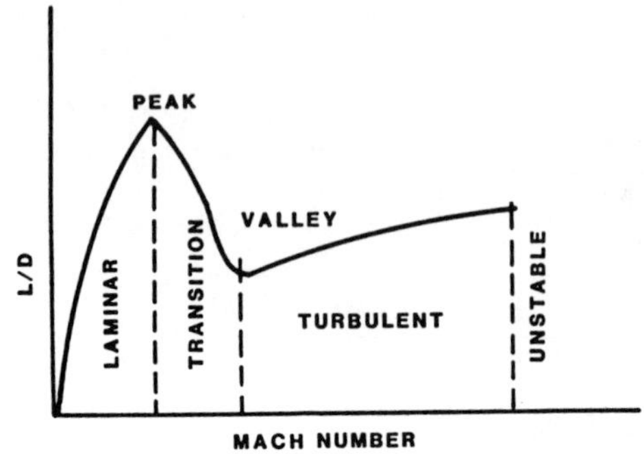

Figure 4-6. Burning characteristics of flames from circular ducts discharging vertically into quiescent air without premixing.

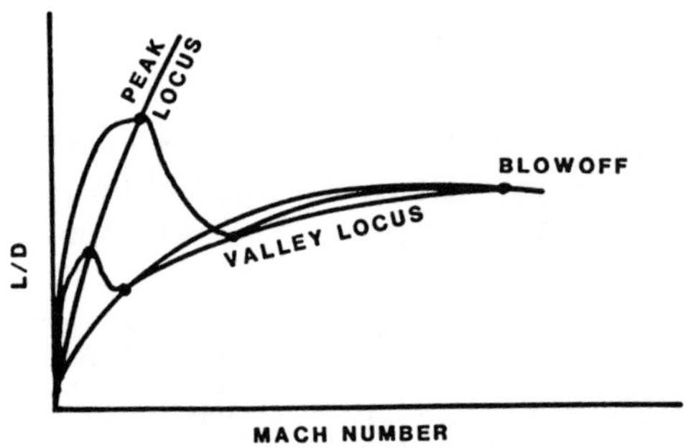

Figure 4-7. Plot of aspect ratio (L/D) versus Mach number.

how a flame profile may be superimposed on the loci of Figure 4-7.

Note that the flame height increases appreciably when combustible gas flow is sufficiently reduced so as to cause a shift back into the laminar zone. By designing a flare tip which induces premixing of gas and air or selecting a "smokeless" design which induces partial premixing by agitation with steam, the in-

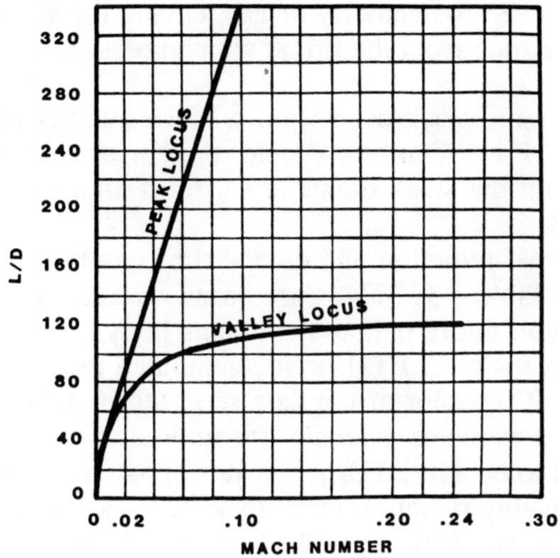

Figure 4-8. Shows superposition of typical flame characteristics on the locus curves.

creased peaking of the flare in the laminar zone may be avoided or materially reduced. This type of flare tip design also reduces the noise level.

Figure 4-8 should be used along with the following criteria:

Peak at Reynolds number	=	3,000
Valley at Reynolds number	=	5,000
Blowoff at Mach number	=	0.2

Note that the Reynolds number is based on stack diameter. Each of these criteria refers to the gas state before combustion at the exit from the stack tip. The Reynolds number of 3,000 applies to the "peak loci curve," the Reynolds number of 5,000 applies to the "valley loci curve," and the blowoff Mach number applies to the limit of the "valley loci curve." The blowoff point is reached when the velocity of gas leaving the stack causes the flame to separate from the tip, at which point the flame becomes unstable.

For maximum stack discharge, a Mach number of 0.2 is recommended. From Figure 4-7 then, the corresponding L/D ratio is 118. From the stack diameter D, the flame length L can be determined.

The thermal radiation and escape time can be estimated from the data in Table 4-1. Values are based on experimental data on the threshold limit of pain to the human body as a function of the radiation intensity in $BTU/hr/ft^2$, generated by a flame.

Figure 4-9 is a plot of the data reported in Table 4-1. A safe level of heat radiation intensity for unlimited time exposure has been found to be 440 $BTU/hr/ft^2$. It is apparent that a time interval with varying radiation intensity must be allowed, to permit a human to escape from a suddenly released intense heat source. The varying radiation intensity results from an individual increasing his distance from the source of heat.

Assume a person is at the base of a flare stack when heat is suddenly released. The average individual reaction time is between three and five seconds. Hence, during this short reaction time interval, the full radiated heat intensity will be absorbed.

Table 4-1
Heat Radiation and Escape Time[12]

Radiation Intensity ($BTU/Hr/Ft^2$)	Time to Pain Threshold (Seconds)
440	∞
550	60
740	40
920	30
1,500	16
2,200	9
3,000	6
3,700	4
6,300	2

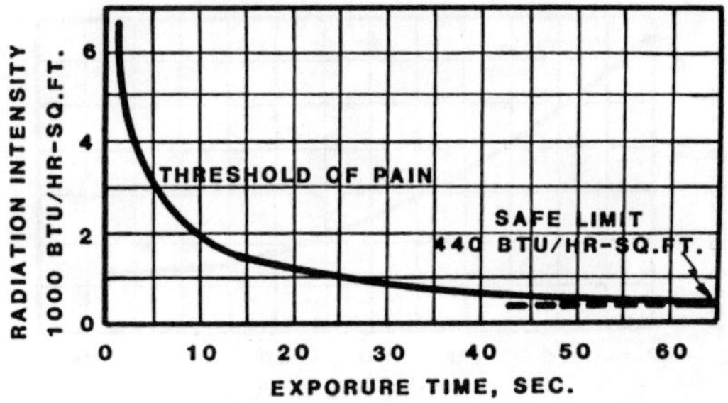

Figure 4-9. Plot of heat radiation vs. exposure time for bare skin at the threshold of pain.

Then follows another short time interval (20 ft/sec is normally assumed to be the average escape velocity of a man) during which continually decreasing amounts of heat will be absorbed until a safe distance is reached (heat intensity for a safe location is 440 BTU/hr/sq ft and lower from Figure 4-9).

The maximum heat intensity that may be tolerated at the base of the stack corresponding to the limiting total heat absorbed may then be determined by the following equation:

$$t_a q_a = t_r q_m + t_e \frac{q_M - q_m}{\ln \dfrac{(q_M)}{(q_m)}} \qquad (4\text{-}6)$$

where $t_a = t_r + t_e$

Total time exposed = reaction time + escape time

$t_a q_a$ = total heat flow/area for the exposure time (Figure 4-9)

q_M = maximum radiation intensity

q_m = minimum radiation intensity

Figure 4-10 is a solution of this equation. The escape time interval t_e depends on the stack height H. Therefore, the use of Figure 4-10 will not be fully understood until the stack height criteria are established.

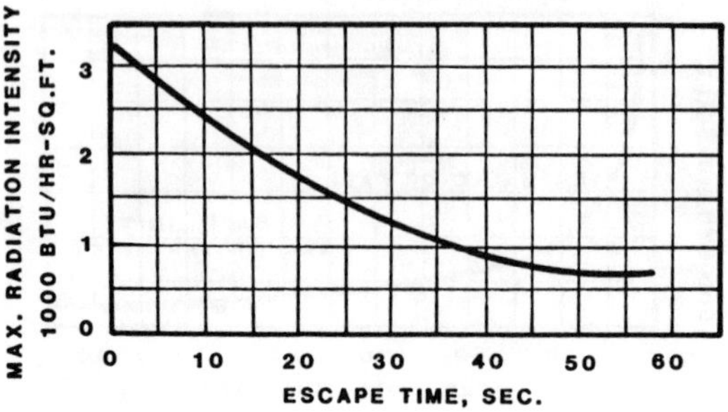

Figure 4-10. Plot of maximum radiation intensity vs. escape time, assuming a 5-second reaction time.

The following steps outline the approach to determining the flare stack height based upon the radiation intensity:

1. Calculate the radiation intensity using the following equation:

$$q = \frac{f_Q}{4\pi X^2} \tag{4-7}$$

where

q = radiation intensity, BTU/hr/sq ft
f = emissivity of the flame
Q = heat generated by the flame, BTU/hr
X = distance from the center of flame, X_m feet above grade to point P, ft (Figure 4-11).

Flame emissivity values for common gases are as follows:[11]

Gas	*f*
Hydrocarbons	0.4
Propane	0.33
Methane	0.2

A relationship between f and the net calorific value of a gas can be used in the absence of data:[11]

$$f = 0.2 \left(\frac{h_c}{900}\right)^{1/2} \qquad (4\text{-}8)$$

where h_c = net heat value of a gas (LHV) in BTU/scf (60°F, 14.7 psia)

2. Calculate the heat flow Q, BTU/hr

$$Q = W \cdot h_c \frac{379}{M} \qquad (4\text{-}9)$$

where

W = lbs/hr of vapors released
h_c = net heating value of gas in BTU/scf (60°F, 14.7 psia)
M = molecular weight of the gas

3. The formula for the stack height is first derived. Referring to Figure 4-11, we have $X^2 = X_m{}^2 + y^2$

and $X_m = \sqrt{H\,(H + L)}$

where

X_m = distance of the point of maximum intensity from grade, ft
H = stack height, ft
L = flame length in ft, L is 118D as shown above

Hence, $X^2 = H\,(H + L) + y^2$

And from Equation 4-7:

$$q_M = \frac{f_Q}{4\pi H\,(H + L)}$$

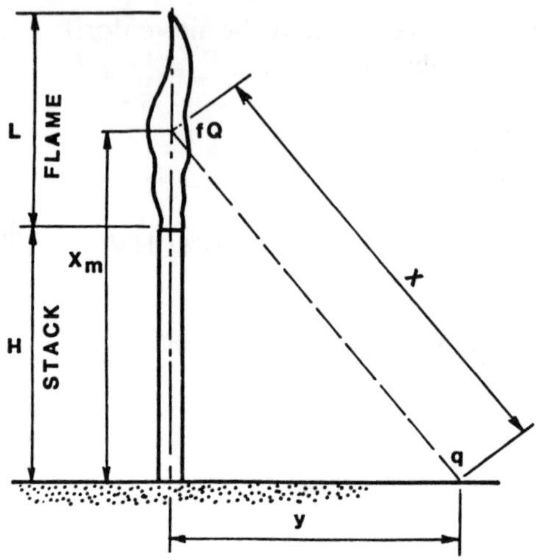

Figure 4-11. Flare stack and flame in stagnant surroundings.

where q_M is the maximum radiation intensity at the base of the flare (i.e., at $y = 0$ or $x = X_m$).

Hence, H is derived as:

$$H = \frac{1}{2}\left[\left(L^2 + \frac{f_Q}{\pi q_m}\right)^{1/2} - L\right] \qquad (4\text{-}11)$$

The shortest stack is obtained when $q_M = 3{,}300$ BTU/hr/sq ft or from Figure 4-10 at $t_e = 0$.

The limiting safe radial distance from the flame is:

$$X = \left(\frac{f_Q}{4\pi 440}\right)^{1/2} \text{ and } X^2 = \frac{f_g}{5530}$$

And we note that y = radial distance from the base of the stack $= [X^2 - H(H + L)]^{1/2}$.

Allowing for the speed of escape (20 ft/sec) we have:

$$y = 20t_e = (X^2 - H\ (H + L)^{1/2} \qquad (4\text{-}12)$$

This defines the safety boundary, corresponding to quiescent ambient air.

Thus, the stack height H, the limiting heat radiation q_M, and the radial distance y can be evaluated with a trial and error procedure by assuming a value of t_e.

The above analysis must be extended to account for the more prevalent case of wind circulation in the vicinity of the flare. For locations where wind intensity is unknown, it is suggested that an average 20 mph wind be assumed in all directions, which results in increasing the safe circular boundary by the resulting tilt of the flame. (This is illustrated in Figure 4-12.) The flame tilt and its effect on the safety boundary increase may be determined as follows:

$$\tan\theta = \frac{U_W}{U} \text{ where } U_W = \text{wind velocity, and}$$

$$U = \text{flare exit velocity.}$$

$$U_W = (X_m - H)\ \text{Sin}\theta \text{ and } U_t = (X_m - H)\ \text{Cos}\theta$$

$$y = \{X^2 - (H + (X_m - H)\ \text{Cos}\theta)^2\}^{1/2} + (X_m - H)\ \text{Sin}\theta \qquad (4\text{-}13)$$

This formula establishes the limiting boundary for wind circulation. When evaluating wind effects on flame tilt, an average wind intensity should be used in the calculations.

Alternate Method of Calculating Stack Height and Safety Boundary

For high flaring rates the stack height calculation previously described leads to very tall stacks. Part of the reason for this conservative estimate is that calculations are based upon the thermal effect on bare skin. If proper clothing is provided to personnel before entering the flare stack area and proper shielding is installed at the stack or at the equipment to reduce the radiation

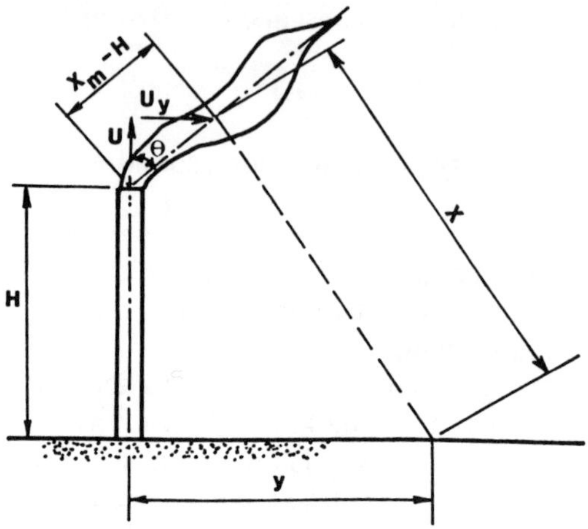

Figure 4-12. Flare stack and flame in wind-blown surroundings.

effects, the required stack height can be greatly reduced. However, there is a tradeoff in that the safe boundary limit must be increased.

Since heat load of the flare, the flame length, and the safe radiation intensity (440 BTU/hr/sq ft) remain the same, decreasing the stack height leads to an increase in the safety radius. Another important consideration is the type of support provided for the stack. This is discussed in the next section. In general, the higher the stack, the greater the structural support costs.

An alternative method of stack height sizing is based on the allowable limit for radiation intensity. For operating personnel, the allowable intensity is 1,500 BTU/hr/sq ft, and for equipment it is 3,000 BTU/hr/sq ft.

The 1,500 BTU/hr/sq ft criterion is established from the following basis. In emergency releases, an operation time of 3 to 5 sec. may be assumed. Perhaps 5 to 10 sec. more would elapse before an individual could escape the area via an average velocity of 20 ft/sec. This would result in a total exposure period ranging

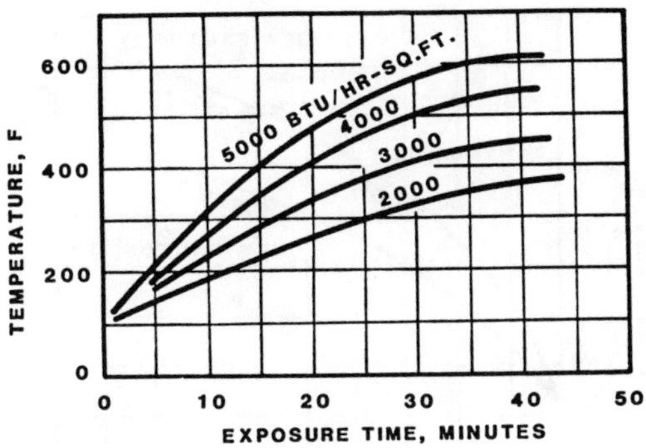

Figure 4-13. Plot of temperature of steel equipment vs. exposure time for different radiant heat intensities. Curves are based on 1/4-in. plate thickness with an effective emissivity of 10 and view factor of 0.5. Cooling caused by convection, etc. are neglected.

from 8 to 15 seconds. The time to pain threshold corresponding to 1,500 BTU/hr/sq ft is 16 seconds from Table 4-1, before the individual could escape to a safe place. The effect of heat radiation on equipment is shown in Figure 4-13.[8] The temperature of metal equipment increases with exposure time and the higher the radiant heat intensity, the greater the temperature. Curve 1 in Figure 4-14 shows the theoretical equilibrium temperatures based on a view factor of 0.5. The actual temperature on surfaces facing the flame will lie between curves 1 and 2.

The temperature of vessels containing liquid or flowing vapors may be lower because of cooling effects. Curve 2 applies to materials having a low heat conductivity coefficient, e.g., wood. In this case, equilibrium temperatures are reached within a shorter time as compared with metal objects. Dehydration of wood takes place at about 500°F, decomposition at 700°F, and ignition at around 800°F, corresponding to 1,300, 3,000 and 4,000 BTU/hr/sq ft, respectively. This means that wooden structures and vegetation exposed to heat intensities of 3,000 to 4,000 BTU/hr/

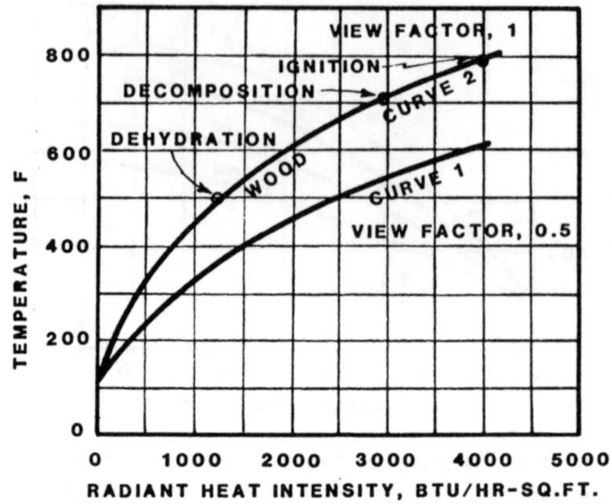

Figure 4-14. Plot of equilibrium temperature vs. radiant heat intensity. Curve 1—metal equipment; Curve 2—wood.

sq ft and higher may catch fire and burn. Paint on equipment also may be damaged. Therefore, it is recommended that equipment located in this area be protected by proper heat shielding, or emergency water sprays.

The following steps outline calculations by the alternate method:

1. (From Equation 4-7) The radial distance from the flame at q = 1500 BTU/hr/sq ft is calculated.
2. The safe radial distance at q = 440 BTU/hr/sq ft is calculated from the same equation.
3. A suitable value for q is assumed at the base of the stack. q = 3000 BTU/hr/sq ft is a good start since protective shielding will be provided in this case at the stack.
4. (From Equation 4-11) H is calculated.

Figure 4-15 illustrates the different heat intensity loci that should be examined. The following sample calculation illustrates these steps.

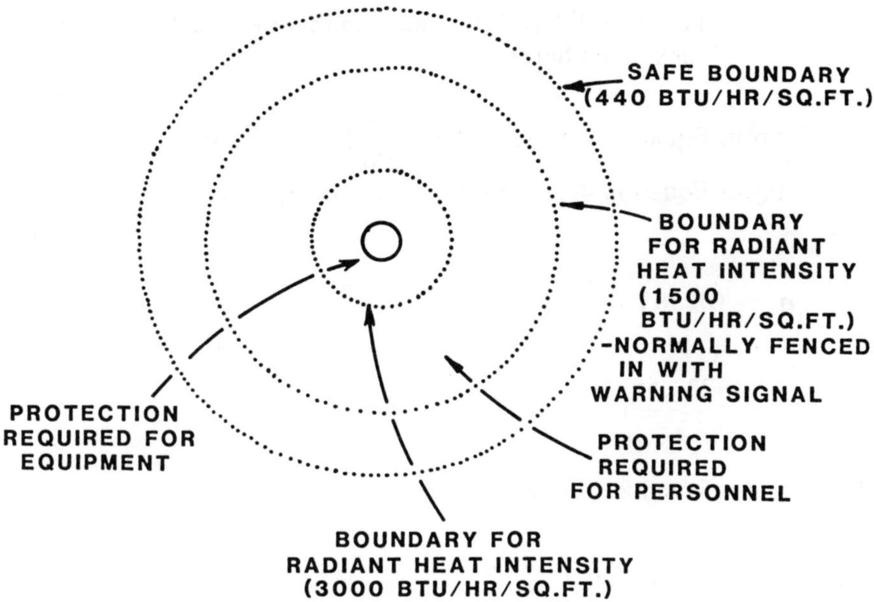

Figure 4-15. Shows contours of radiant heat intensity.

Example 4-3

Given the same data as in Example 4-2 (1,000,000 lbs/hr maximum load, 800,000 lbs/hr normal load, M.W. = 50, stack diameter = 48 in., temperature = 500°F of the relieving vapor and average wind velocity = 20 miles per hr, net heating value = 1500 BTU/scf) calculate the stack height and the safe boundary.

Solution:

The total heat released is calculated from Equation 4-9:

$$Q = \frac{1,000,000 \times 379}{50} = 1500 \text{ BTU/hr/ft}^2$$

$= 11{,}370 \times 10^6$ BTU/hr (maximum vapor load should be considered here)

From Equation 4-8: $f = 0.2 \left(\frac{1500}{900}\right)^{1/2} = 0.258$

From Equation 4-7: the radiant heat intensity is:

$$q = 0.258 \left(\frac{11{,}370}{4\pi X^2}\right) 10^6$$

$$= \left(\frac{233.55}{X^2}\right) 10^6$$

The safe radial distance is:

$$X = \sqrt{\left(\frac{233.55}{440}\right) 10^6}$$

$$= 728.55 \text{ ft.}$$

From Figure 4-7, for a Mach number of 0.2, the flame length is obtained:

$L = 118D$
$= 480$ ft; where D = stack diameter in ft

From Equation 4-11: the stack height is

$$H = \frac{1}{2}\left[\left(L^2 + \frac{f_Q}{\pi q_m}\right)^{1/2} - L\right] \tag{4-11}$$

The shortest stack height using a zero escape velocity is computed from the following:

$q_M = 3300$ BTU/hr/sq ft from Figure 4-1
$H = 118.25 \simeq 120$ ft.

Obviously, this value cannot be used because of safety reasons. Hence, the height must be calculated assuming some reasonable escape time.

Assuming

$$t_e = 30 \text{ sec}$$
$$q_M = 1330 \text{ BTU/hr/sq ft}$$
$$H = 248 \text{ ft}$$

Calculated escape time = 29.5 secs, which is in reasonable agreement with the assumed value.

Hence, 248 ft is the selected flare stack height.

We now calculate the average wind effect on the safe boundary around the flare stacks.

$$\text{Wind velocity} = 20 \text{ miles an hour} = 29.3 \text{ ft/sec}$$

$$\text{Flame tilt angle, } \text{Tan}\theta = \frac{29.3}{245} = 0.1196$$

where the flare exit velocity is 245 ft/sec based on the maximum load.

$$\text{Sin}\theta = 0.1187 \text{ and } \text{Cos}\theta = 0.99, X_m = H(H + L) = \sqrt{425} \text{ ft}$$

$$\text{Hence, } X_m - H = 425 - 248 = 177 \text{ ft}$$
$$(H + (X_m - H)\text{ Cos}\theta)^2 = (248 + 177 \times 0.99)^2 = 179{,}123$$

$$\text{And } X^2 = (728.55)^2 = 530{,}785.1$$

Hence, safe radial distance from the base of the stack:

$$y = (530{,}785 - 179{,}123)^{1/2} + 177 \times 0.1187 = 572 \text{ ft}$$

Estimating Ground Level Concentrations

After the stack height has been established from radiation intensity values, the maximum permissible ground level concentration of toxic gases in the event of a flame blowout should be evaluated. Table 4-2 represents the toxicological threshold limits as allowed by the Environmental Protection Agency (EPA).

Table 4-2
Threshold Limits for Certain Toxic Substances
Gases and Vapors

Gas or Vapor	PPM
Acetaldehyde	200
Acetic acid	10
Acetic anhydride	5
Acetone	1,000
Acrolein	0.5
Acrylonitire	20
Ammonia	100
Amyl acetate	200
Amyl alcohol	100
Aniline	5
Arsinic	0.05
Benzene	35
Benzyl chloride	1
Bromide	1
Butadiene	1,000
Butyl alcohol	100
Butylamine	5
Carbon dioxide	5,000
Carbon disulfide	20
Carbon monoxide	100
Carbon tetrachloride	25
Chlorine	1
Chlorobenzene	75
Chloroform	100
Cresol (all isomers)	5
Cyclohexane	400
Cyclohexanol	100
Cyclohexanone	100
Cyclohexene	400
Cyclopropane	400
Diacetone alcohol	50
0-Dichlorobenzene	50
1,1-Dichloroethane	100
Diethylamine	25
Diisobutyl ketone	50
Dimethylaniline	5

Table 4-2. Continued.

Gas or Vapor	PPM
Dimethylsulfate	1
Diethylene dioxide	100
Ethyl acetate	400
Ethyl alcohol (ethanol)	1,000
Ethylamine	25
Ethylbenzene	200
Ethyl bromide	200
Ethyl chloride	1,000
Ethyl ether	400
Ethylene chlorohydrin	5
Ethylenedriamine	10
Ethylene dibromide	25
Ethylene dichloride	100
Ethylene oxide	100
Fluorine	0.1
Formaldehyde	5
Gasoline	500
Hydrazine	1
Hydrogen selenide	0.05
Hydrogen sulfide	20
Isodine	0.1
Isophorene	25
Isopropylamine	5
Mesityl oxide	50
Methyl acetate	200
Methyl acetylene	1,000
Methyl alcohol	200
Methyl bromide	20
2-Methoxyethanol	25
Methyl chloride	100
Methylcyclohexane	500
Methylcyclohexanol	100
Methylcyclohexanone	100
Methyl formate	100
Methyl amyl alcohol	25

(Continued)

Table 4-2. Continued.

Gas or Vapor	PPM
Methylene chloride (dichloromethane)	500
Naphtha (coal tar)	200
Naphtha (petroleum)	500
pNitroaniline	1
Nickel carbonyl	0.001
Nitrobenzene	1
Nitroethane	100
Nitrogen dioxide	5
Nitromethane	100
Nitrotoluene	5
Octane	500
Ozone	0.1
Pentane	1,000
Propyl ketone	200
Phenol	5
Phenylhydrazine	5
Phosgene (carbonyl chloride)	1
Phosphine	0.05
Phosphorus trichloride	0.5
Propyl acetate	200
Propyl alcohol	400
Propyl ether	500
Propylene dichloride	75
Pyridine	10
Quinone	0.1
Stibine	0.1
Styrene	200
Sulfur dioxide	10
Sulfur hexafluoride	1,000
Sulfur monochloride	1
Sulfur pentafluoride	0.025
1,1,2,2-Tetrachloroethane	5
Tetranitromethane	1
Toluene (toluol)	200
o-Toluidine	5
Trichloroethylene	200

Estimated ground level concentrations should be based on the emergency condition of flame blowout. The calculation is normally done for a range of climatological conditions at the plant site. For a rough estimate, the following empirical formula[13] may be used:

$$C_{max} = \frac{3{,}697\ VMD_Z}{\mu H^2 D_y} \qquad (4\text{-}14)$$

$$X_{max} = (H/D_Z)^{2/(2-N)}$$

where

C_{max} = concentration at grade in ppm (volume)
V = sp. vol. of toxic gas, cu ft per lb
M = weight discharge of pollutant component, in tons per day
D_Z = vertical diffusion coefficient
μ = air velocity at grade
H = stack height, ft
D_y = horizontal diffusion coefficient
X = distance from stack to the point of maximum conc., ft
N = environmental factor

The following values are taken from the API Manual:[14]

D_Z = 0.13
D_y = 0.13
N = 0.25

Cheremisinoff et al.[15,16] give detailed procedures for estimating pollution levels. The following example illustrates ground level concentration estimation.

Example 4-4

Assume 1,000 lbs per hr hydrogen sulfide in the vent stream (when the flare is extinguished). The mean wind velocity at the plant site is 5 mph, the stack height is 248 ft, and the temperature of the relieving gas is 300°F. Determine the maximum allowable ground level concentration of H_2S and the distance where it occurs.

Solution:

$$V = \frac{379}{34} \times \frac{760}{520} = 16.29 \text{ cu. ft per lb}$$

$$M = 12 \text{ tons per day}$$

$$D_z = 0.13,\ D_y = 0.13,\ \mu = 5 \text{ mph}$$

$$H = 248 \text{ ft},\ N = 0.25$$

$$C_{max} = 5.792 \text{ ppm (volume)}$$

$$X_{max} = 4875 \text{ ft}$$

From Table 4-2, the threshold limit for H_2S is 20 ppm. Hence, C_{max} is well below the allowable limit, and the final stack height is 248 ft.

Stack Support

There are generally three types of flare stack supports: Guyed type, Derrick, and Self-supporting.

As a rough guide to the economics of these three types of flare structures, the comparative costs for material and labor as functions of stack height are tabulated here:

Capital Investment (Equipment Only)

	Up to 150 Ft	**150 Up to 200 Ft**	**Above 200 Ft**
Least expensive	Derrick Type	Derrick Type	Guyed
	Self-supporting	Guyed	Derrick
Most expensive	Guyed	Self-supporting	Self-supporting
Installation Labor			
Least expensive	Self-supporting	Derrick	Guyed
	Guyed	(Self-supporting	Derrick
	Derrick	Guyed)*	Self-supporting
Most expensive			

*about equal in cost

To ensure ignition of flare gases, continuous pilots with a means of remote ignition are recommended for all flares. Generally, the pilot system consists of three components—a continuous pilot, an on/off pilot, and an igniter. The most commonly used type of igniter is the flame-front propagation type which utilizes a spark from a remote location to ignite a flammable mixture. The on/off type is used only to ensure ignition of the continuous pilot. Pilot igniter controls are located near the base of elevated flares and at least 100 ft from ground flares.

The number of pilot systems required per flare is largely a function of wind conditions. A minimum of two pilot systems is recommended with normally three pilot systems used. These are distributed uniformly around the top of the flare.

A typical flare pilot system for an elevated flare stack is shown in Figure 4-16.[17] The same type of assembly, installed horizon-

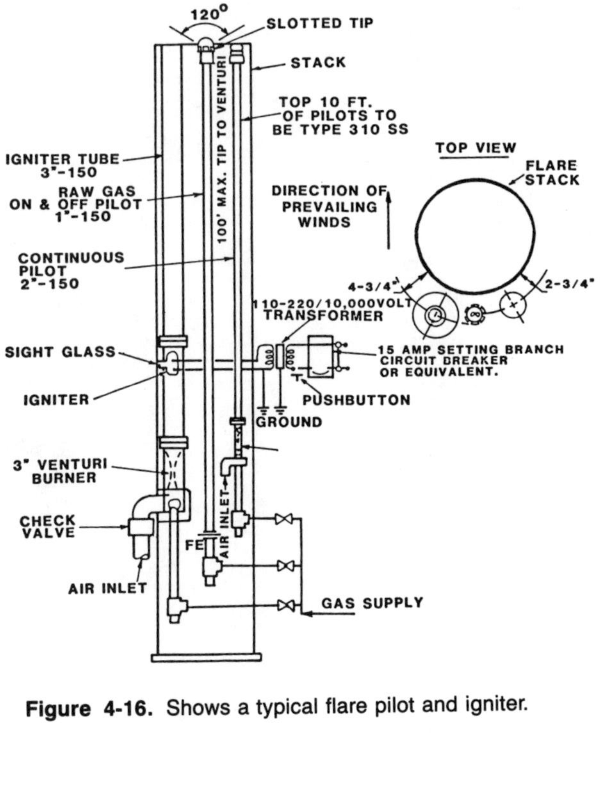

Figure 4-16. Shows a typical flare pilot and igniter.

tally, may be used for ground flares. The following description applies to each pilot system.

The pilot proper is piped to the top of the flare stack via a 2-in. Venturi burner. Nozzles are provided at the end of the pipe. In some designs, nozzles are hooded and should the flame blow out, the heat of the nozzle will immediately reignite it.

In the pilot igniter system, the gas pipe is connected to a 3 in. Venturi type burner, which is located at the bottom of the stack. The fuel gas flows through a nozzle to inspirate air to form a combustible mixture. The igniter with spark gap is located approximately 3 ft above the burner. When the igniter button is pushed, the resulting spark ignites the gas-air mixture. The flame front generated travels up the pipe at the top of the flare and ignites the gas from the pilot nozzles.

5

FINAL DESIGN CONSIDERATIONS

Materials of Construction

The following table outlines materials of construction for different components of the flare system.

Component		Material of Construction
Piping &	Up to −20°F	Conventional carbon steel
Knockout Drum	Up to −50°F	Special low temp carbon steel
	−150°F & below	18-8 stainless steel
	Above 750°F	High temperature resistant alloy

Stack

Bottom Section	Gunite line (cemented for corrosion resistance)
Burner tips (about 10 ft)	Stainless steel line with refractories
Section up to 20 ft below burner tips	High temperature resistant refractories
Other sections of the stack	Special low temperature carbon steel
Structural members, hardware and boltings	Should be hot dip galvanized after fabrication

Steam and Fuel Requirements

The following empirical formula[18] is recommended for evaluating the requirement of steam for producing a smokeless flame as a function of the flow rate of hydrocarbons and their molecular weight:

$$W_s = W_H \left(0.68 - \frac{10.8}{M}\right) \tag{5-1}$$

where W_s = steam, lbs/hr
W_H = hydrocarbons, lbs/hr
M = molecular weight

It may be observed from this that the higher the molecular weight, the higher the required steam consumption. This may be associated with the theory that the higher the molecular weight, the lower the ratio of steam to carbon dioxide after combustion, resulting in a greater tendency to smoke (see Figure 5.1).

Since steam consumption is rather high (about 0.464 lb per lb of hydrocarbon for mol wt = 50), it is too expensive to provide for smokeless burning for the maximum flare load. Normally, 20 percent of the maximum load is designed for smokeless burning. This is well supported by the fact that massive failure is very rare and in 90 percent of the occurrences smokeless flares are produced.

Fuel gas supply to the pilots and igniters must have a high reliability. Since normal plant fuel sources may be upset or lost, it is desirable to provide a back-up system connected to the most reliable alternate fuel source with provision for automatic cut-in on low pressure. The flare fuel system should be carefully checked to ensure that hydrates cannot present a problem. Because of small lines, long exposed runs, and large vertical rises up the stack, use of a liquid knock-out pot is frequently warranted to remove condensates that may have collected in the fuel line, especially during winter. It is good practice to provide a low-pressure alarm on the fuel supply after the last regulator, which will warn the operator.

Figure 5-1. (A) Steam flare in a chemical plant—steam off. (B) Smokeless steam flare—steam on. (Courtesy: National Air Oil Burner Co., Philadelphia, PA).

For cases where the discharged vapors have a heating value of less than 150 BTU/SCF, a supply of fuel gas is injected into the flare header, thereby raising its heating value to acceptable limits. Sensors are provided in the effluent from relief valves in these streams to actuate an alarm in the control room signalling the need for supplemental fuel gas.

Purging of Flare Lines

Any gas or mixture of gases that cannot reach dew point at any condition of ambient temperature can be used as a purge for flare systems. Nitrogen, methane, or natural gas are normally used as purge gases.

Purging is normally of two types: normal purging and emergency purging.

Normal purging is used continuously and admitted to the flare system at the end of each subheader and at the bottom of the molecular seal at the flare stack. A normal recommended purge volume, when the molecular seal is used, is that purge volume which will create a velocity of 0.1 ft per sec at the flare tip. When a molecular seal is not provided, the exit velocity is 1 ft/sec. The purge volume depends upon the wind velocity at the flare elevation. These velocity criteria are based on a wind velocity of 15 mph and vary as the square of the wind velocity.

Emergency purging is used to compensate for thermal shrinkage. After cessation of hot vent gas flow, the system residual gas will shrink as it cools to the ambient temperature. It normally takes about 15 minutes to reach equilibrium. Unless the purge gas is admitted to the system, the shrink will draw air back into the flare header. The shrink problem can be overcome by sensing the system temperature and adding make-up gas at a rate commensurate with the system volume and the maximum anticipated gas temperature.

Noise Pollution

Noise pollution from flares has for too long been an inconvenience, accepted in petrochemical plants as an inevitable by-product of the flaring process. It has been established that the major individual source of noise from an elevated flare is usually at the flare tip itself. This is especially true when the flare tip is of the type used for smokeless flaring of hydrocarbon gases utilizing steam injection. Basically, noise is created because of two reasons: steam energy losses at the high pressure steam injectors, and unsteadiness in the combustion process.

Ground flares are normally quieter than elevated flares. This is probably due to the fact that the flame contained inside a box is protected from wind effects and the stabilizing effect of the heat reradiated from the refractory walls reduces the random characteristics of combustion. The walls themselves will absorb some of the sound energy.

Sophisticated designs of flare tips have greatly reduced the noise pollution. In some designs, combustion efficiency has been greatly increased by remixing of air with gas before they are combusted. Steam is also premixed with air and gas before gases leave the flare tip. Some of the turbulent noise energy is thus shielded by the tip itself.

Stress Relief, Winterizing, and Controls

The major *stresses* to which the discharge piping of a relief system is subject are results of thermal strains from entry of cold or hot gases. Temperature fluctuations are normally very wide. In the majority of situations it is usually possible to maintain stress levels within allowable limits over the full temperature range by providing an expansion joint or expansion with a cold or hot spring. Special attention to stresses is recommended where piping constructed of carbon steel is used for metal temperatures as low as −50°F.

Design of discharge piping requires careful analysis of the possible thermal and mechanical stresses imposed on the pressure-relief valves. Proper anchors, supports and provision for flexibility of discharge piping can prevent these stresses. These are discussed in detail in Reference 19.

Winterizing of the flare system depends upon the severity of ambient temperatures. It is a normal practice to slope the flare headers toward the knock-out drum 1/4 in. per 10 ft of run. This enables condensate to flow back into the knock-out drum, thereby reducing the possibility of a pipe freeze up due to lengthy exposure to low ambient temperature. Knock-out drums are usually provided with a submerged steam heater in order to prevent freezing. Where a water seal is used, a similar arrangement is warranted. In some cold climate areas, flare headers containing water are steam traced and insulated.

Instrumentation normally used in flare systems was summarized in Figure 1-3. The major areas requiring instrumentation and controls are as follows:

1. To ensure smokeless burning, a suitable control system is provided to regulate steam injection into the flare tip. Normally a flow sensor is provided on the main flare header. The flow sensor is designed to pass, with a predetermined back pressure, the maximum smokeless rate of the flare. This is in ratio control with steam flow.
2. Thermocouples are provided for the pilots connected to an alarm in the control room.
3. An oxygen analyzer with an alarm is normally provided to indicate the presence of the air or oxygen in the flare system.
4. The knock-out drum is level-controlled in order to maintain a constant level for providing a seal and to prevent the pump from running dry.

Startup and Shutdown Procedures

These general guidelines should be followed when starting up and shutting down a flare system. An initial *system checkout* is first needed:

1. After the completion of construction, the system should be thoroughly flushed with water to remove scale and debris. Pressure testing should be conducted where required. Special attention should be given to all flanged joints, valves, and connections. All leaks found should be repaired and retested.

2. The motor drive slop pump and flare knock-out drum pump should be checked for ease of operation and correct rotation.
3. All instruments should be checked for proper connections and performance.
4. Equipment such as flare tip, molecular seal, flame front generator, flow sensor, and all associated piping should be given a final check.

System Purge: The flare system must be purged of air before the pilots are ignited, otherwise there is danger of a severe explosion. After the flare system has been purged of air (less than 2 percent oxygen) as recorded from the oxygen analyzer, the pilots are lighted as follows:

1. All valves in the flame front generator are closed.
2. Plant air and fuel gas lines up to the flame front generator should be blown down to remove any line condensate before gas or air is admitted.
3. Push the ignition button and check for a spark at the sight port.
4. Open valves for the flame front generator to pilot No. 1 and fuel gas to all pilots.
5. Open the gas supply to approximately 10 psig by observing the pressure gauges.
6. Purge for three minutes. Then push igniter button to light the pilot. Then push light pilot No. 2 and No. 3 in the same manner.

Shutdown: The total flare system can only be shut down and isolated after all the process units are shut down, drained of hydrocarbons, depressurized, and purged as necessary. Then the flare system is purged with nitrogen before opening up the knock-out drum, molecular seal, etc., for any maintenance.

Individual process units or pieces of equipment can be isolated from the operating flare system after they are shut down by closing block valves and installing blind flanges, when maintenance is required. Further guidelines are given in Reference 20.

6

GENERAL FLOW CALCULATION NOTES AND UNIT CONVERSIONS

Basic Definitions and Properties of Fluids

Mass density of a fluid:

$$\rho' = \text{Mass per unit volume} = \frac{\text{Mass}}{\text{Unit Volume}} = \frac{w}{gv}$$

w–lb
g–ft/sec^2
v–ft^3

$$\rho' = \frac{\text{lb}}{\text{ft-sec}^{-2}(\text{ft}^3)} = \frac{\text{lb-sec}^2}{\text{ft}^4} = \text{lb sec}^2/\text{ft}^4$$

Specific weight of a fluid:

$$\rho = \text{Weight per unit volume} = \frac{\text{Weight}}{\text{Unit Volume}} = \frac{w}{v}$$

$$\rho = \frac{\text{lb}}{\text{ft}^3}$$

Specific gravity:

γ = Weight of a substance compared with weight of an equal volume of water

$$\gamma = \frac{\rho}{\rho_w}$$

where ρ = density of substance
ρ_w = density of water 62.4 lb/ft^3

Viscosity:

Coefficient of viscosity—the resistance offered by, or the drag transmitted through the fluid by a layer of fluid of unit area to the motion parallel to this area of another layer of fluid, at unit distance, moving with unit velocity relative to the first layer.

Dynamic viscosity—indicates the force needed to displace a layer 1 cm^2 in area at a velocity of 1 cm/sec at a distance of 1 cm from a static layer of equal size.

Unit measurement is poise (p) or centipoise (cp). Common symbols μ or η.

1 $kgfs/m^2$ = 98.1 p = 9810 cp.

Kinematic viscosity—ratio of viscosity to density of fluid. Unit of measurement is Stokes (S) or centistokes (cS). Common symbol ν.

1 cm^2/sec = 1 S = 100 cS.

$$\nu = \frac{\mu}{\rho}$$

Hydraulics: The science of hydraulics is divided into two subjects.

1. Hydrostatics, which is the theory of equilibrium of fluids.
2. Hydrodynamics, which is the theory of the motion of fluids under the influence of force.

Principles of Hydrostatics

Pascal's Law—If pressure is exerted upon a mass of liquid:

- It is transmitted with equal intensity in all directions.
- It acts with the same force on all equal areas.
- It acts in a direction at right angles to those areas.

Hydrostatic Pressure: Refer to Figure 6-1.

$$P = \frac{F_1}{A_1} = \frac{F_2}{A_2},$$

A_1, A_2–area
F_1, F_2–force

Piston Forces: Refer to Figure 4-1.

$$F_1 = PA_1 = F_2 \frac{A_1}{A_2}$$

$$F_2 = PA_2 = F_1 \frac{A_2}{A_1}$$

Buoyancy: Archimedes principle states that the buoyancy F_A of a body is equal to the weight of the water displaced.

$F_A = \gamma V$, γ–specific gravity
V–submerged or displaced volume

when F_A is less than the weight of the body, it will sink.
when F_A is equal to the weight of the body, it will remain suspended.
when F_A is greater than the weight of the body it will float.

Atmospheric pressure: the pressure of the air and the surrounding atmosphere.
Barometric pressure: the same as atmospheric pressure.
Absolute pressure: a measure of pressure referenced to a complete vacuum, or zero pressure.

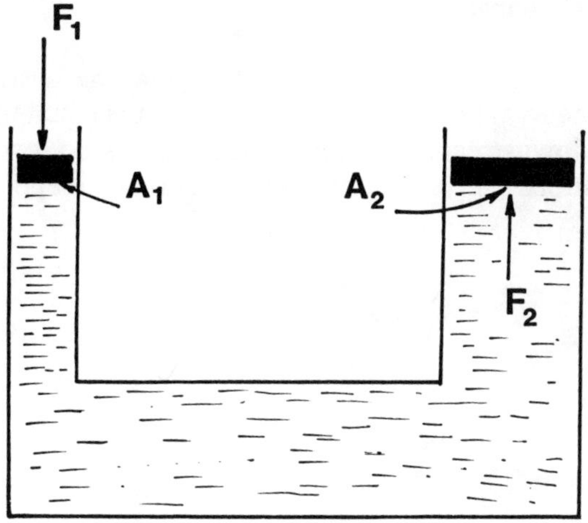

Figure 6-1. Illustrates hydrostatic pressure and piston forces.

Gauge pressure: pressure expressed as a quantity measured from (above) atmospheric pressure or some reference pressure.
Vacuum: method of expressing pressure as a quantity below atmospheric or some reference pressure.

Principles of Hydrodynamics

Equation of Continuity (Conservation of Mass)—For steady flow, both uniform and nonuniform in any continuous stream of liquid, the discharge passing through any cross section is constant.

$Q = AV$, or for different cross sections along a stream—
$Q = A_2V_2 = A_3V_3$

where Q = volume of flow per unit time (ft^3/sec)
A = cross-sectional area of stream (ft^2)
V = average velocity of stream (ft/sec)

Bernoulli's Theorem

Comparison of conditions at two points along a stream line shows that the total head at the upstream point is equal to the total head at the downstream point, provided there is no loss between the two positions due to friction and no gain due to application of outside work.

$$\frac{V_1^2}{2g} + \frac{P_1}{\rho} + Z_1 = \frac{V_2^2}{2g} + \frac{P_2}{\rho} + Z_2$$

$$V = \sqrt{2gh}, \quad h = \frac{V^2}{2g}$$

note

$\frac{V^2}{2g}$ is the velocity head (ft)

Z is the potential head (ft)

$\frac{P}{\rho}$ is the pressure head (ft of fluid)

To include loss or gain, refer to Figure 6-2 and use following:

$$\frac{V^2}{2g} + \frac{P_1}{\rho} + Z_1 + H_A - H_E - H_L = \frac{V_2^2}{2g} + \frac{P_2}{\rho} + Z_2$$

H_E is extracted head (ft)
H_L is lost head (ft)
H_A is added head (ft)

$$\text{Fluid horsepower (HP) at a pump shaft} = \frac{Q\rho H}{550}$$

where H = total head, ft (1 H.P. = 550 ft-lb/sec).

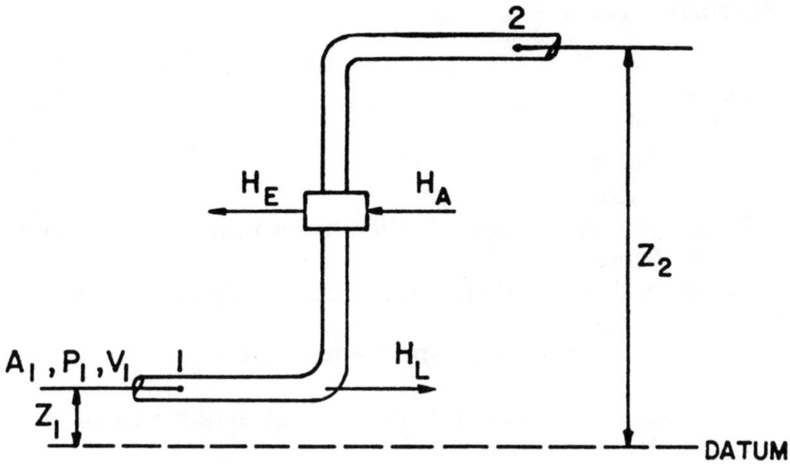

Figure 6-2. To account for loss or gains in flow system.

$$\text{H.P. at pump motor} = \frac{Q\rho H}{550} \times \frac{1}{\eta}$$

where η = efficiency and Q is volumetric flow in ft^3/sec.

Flow in Pipe and Fittings:
Friction loss in pipes—

$$H_f = f \frac{L}{D} \frac{V^2}{2g}$$

where L = pipe length (ft)
D = inside diameter of pipe (ft)
V = fluid velocity in pipe (ft/sec)
f = friction factor
g = gravity (ft/s^2)

Reynolds Number:

$$\mathrm{Re} = \frac{DV\rho}{\mu}$$

ρ = density (lb/ft^3)
μ = viscosity ($lb\text{-}sec/ft^2$)
for Re < 2,000 compute friction factor from (laminar flow):
$f = 64/\mathrm{Re}$
for Re > 4,000 (turbulent flow) use Moody Plot or other.

Gas Laws and Nomenclature

Pressure, P, is measured in atmospheres, millimeters of mercury (mmHg), or Torricellis (Torr).

Volume, V, is measured in cubic centimeters (cm^3), milliliters (ml), or liters.[1]

The number of moles, n, is calculable as the weight of sample divided by its molecular weight or molar weight. The number of moles can also be expressed as the number of molecules divided by Avogadro's number 6.02×10^{23}.

Temperature, T, is measured in degrees absolute (°A) or degrees Kelvin (°K). Temperature in degrees Celsius or centigrade (°C) can be converted to absolute temperature by adding 273.15°.

$$T°K = t°C + 273.15°$$

$$t°C = \frac{5}{9}(t°F - 32)$$

Boyle's Law—For an ideal gas the volume of a fixed weight at a fixed temperature varies inversely with the pressure exerted on it.

Charles' Law—Given an ideal gas, its volume is directly proportional to the absolute temperature, provided that the pressure remains constant and the gas sample is of fixed weight.

Dalton's Law of Partial Pressures—For more than one gas in a fixed volume container, the total pressure exerted by the mixture is the sum of the pressures that would have been exerted if each gas were there alone. Partial pressure is the individual pressure one component exerts, treating it as if it were all by itself in the container. Dalton's law states that the observed pressure is the sum of the partial pressures:

$$P = P_1 + P_2 + P_3 + \ldots$$

Standard Conditions—the reference conditions for comparison, 0°C and 1 atm. pressure. Referred to as STP (standard temperature and pressure).

Avogadro's Principle—At the same pressure and temperature, equal volumes of gases contain equal numbers of particles. One mole of any gas contains the Avogadro number of molecules (6.02×10^{23} molecules). At STP, the volume occupied by 6.02×10^{23} molecules is equal to 22.4 liters.

Molar Volume of an ideal gas—At STP one mole of gas occupies 22.4 liters.

Equation of State for Ideal Gases: $PV = nRT$
where P is pressure, V is volume, n is the number of moles, T is absolute temperature and R is the universal gas constant. Table 6-1 gives values of R in different units.

Table 6-1
Universal Gas Law Constants

Type Unit	Value	Units
Mechanical	0.082054	liter-atm-mole^{-1}–deg k^{-1}
Mechanical	82.054	ml-atm-mole^{-1}–deg k^{-1}
cgs	8.3144×10^7	ergs-mole^{-1}–deg k^{-1}
Electrical	8.3144	joules-mole^{-1}–deg k^{-1}
Thermal	1.9872	calories-mole^{-1} deg k^{-1}

Conversion Factors*

TO CONVERT	INTO	MULTIPLY BY
	A	
Abcoulomb	Statcoulombs	2.998×10^{10}
Acre	Sq. chain (Gunters)	10
Acre	Rods	160
Acre	Square links (Gunters)	1×10^{5}
Acre	Hectare or sq. hectometer	.4047
acres	sq feet	43,560.0
acres	sq meters	4,047.
acres	sq miles	1.562×10^{-3}
acres	sq yards	4,840.
acre-feet	cu feet	43,560.0
acre-feet	gallons	3.259×10^{5}
amperes/sq cm	amps/sq in.	6.452
amperes/sq cm	amps/sq meter	10^{4}
amperes/sq in.	amps/sq cm	0.1550
amperes/sq in.	amps/sq meter	1,550.0
amperes/sq meter	amps/sq cm	10^{-4}
amperes/sq meter	amps/sq in.	6.452×10^{-4}
ampere-hours	coulombs	3,600.0
ampere-hours	faradays	0.03731
ampere-turns	gilberts	1.257
ampere-turns/cm	amp-turns/in.	2.540
ampere-turns/cm	amp-turns/meter	100.0
ampere-turns/cm	gilberts/cm	1.257
ampere-turns/in.	amp-turns/cm	0.3937
ampere-turns/in.	amp-turns/meter	39.37
ampere-turns/in.	gilberts/cm	0.4950
ampere-turns/meter	amp/turns/cm	0.01
ampere-turns/meter	amp-turns/in.	0.0254
ampere-turns/meter	gilberts/cm	0.01257
Angstrom unit	Inch	3937×10^{-9}
Angstrom unit	Meter	1×10^{-10}
Angstrom unit	Micron or (Mu)	1×10^{-4}
Are	Acre (US)	.02471
Ares	sq. yards	119.60
ares	acres	0.02471
ares	sq meters	100.0
Astronomical Unit	Kilometers	1.495×10^{8}
Atmospheres	Ton/sq. inch	.007348
atmospheres	cms of mercury	76.0
atmospheres	ft of water (at 4°C)	33.90
atmospheres	in. of mercury (at 0°C)	29.92
atmospheres	kgs/sq cm	1.0333
atmospheres	kgs/sq meter	10,332.

* From *Applied Process Design for Chemical and Petrochemical Plants*, Vols. 1–3, 2nd ed. by E. E. Ludwig, Gulf Publishing Company.

TO CONVERT	INTO	MULTIPLY BY
atmospheres	pounds/sq in.	14.70
atmospheres	tons/sq ft	1.058
	B	
Barrels (U.S., dry)	cu. inches	7056.
Barrels (U.S., dry)	quarts (dry)	105.0
Barrels (U.S., liquid)	gallons	31.5
barrels (oil)	gallons (oil)	42.0
bars	atmospheres	0.9869
bars	dynes/sq cm	10^6
bars	kgs/sq meter	1.020×10^4
bars	pounds/sq ft	2,089.
bars	pounds/sq in.	14.50
Baryl	Dyne/sq. cm.	1.000
Bolt (US Cloth)	Meters	36.576
BTU	Liter—Atmosphere	10.409
Btu	ergs	1.0550×10^{10}
Btu	foot-lbs	778.3
Btu	gram-calories	252.0
Btu	horsepower-hrs	3.931×10^{-4}
Btu	joules	1,054.8
Btu	kilogram-calories	0.2520
Btu	kilogram-meters	107.5
Btu	kilowatt-hrs	2.928×10^{-4}
Btu/hr	foot-pounds/sec	0.2162
Btu/hr	gram-cal/sec	0.0700
Btu/hr	horsepower-hrs	3.929×10^{-4}
Btu/hr	watts	0.2931
Btu/min	foot-lbs/sec	12.96
Btu/min	horsepower	0.02356
Btu/min	kilowatts	0.01757
Btu/min	watts	17.57
Btu/sq ft/min	watts/sq in.	0.1221
Bucket (Br. dry)	Cubic Cm.	1.818×10^4
bushels	cu ft	1.2445
bushels	cu in.	2,150.4
bushels	cu meters	0.03524
bushels	liters	35.24
bushels	pecks	4.0
bushels	pints (dry)	64.0
bushels	quarts (dry)	32.0
	C	
Calories, gram (mean)	B.T.U. (mean)	3.9685×10^{-3}
Candle/sq. cm	Lamberts	3.142
Candle/sq. inch	Lamberts	.4870
centares (centiares)	sq meters	1.0
Centigrade	Fahrenheit	$(C° \times 9/5)+32$
centigrams	grams	0.01
Centiliter	Ounce fluid (US)	.3382
Centiliter	Cubic inch	.6103

(Continued)

TO CONVERT	INTO	MULTIPLY BY
Centiliter	drams	2.705
centiliters	liters	0.01
centimeters	feet	3.281×10^{-2}
centimeters	inches	0.3937
centimeters	kilometers	10^{-5}
centimeters	meters	0.01
centimeters	miles	6.214×10^{-6}
centimeters	millimeters	10.0
centimeters	mils	393.7
centimeters	yards	1.094×10^{-2}
centimeter-dynes	cm-grams	1.020×10^{-3}
centimeter-dynes	meter-kgs	1.020×10^{-8}
centimeter-dynes	pound-feet	7.376×10^{-8}
centimeter-grams	cm-dynes	980.7
centimeter-grams	meter-kgs	10^{-5}
centimeter-grams	pound-feet	7.233×10^{-5}
centimeters of mercury	atmospheres	0.01316
centimeters of mercury	feet of water	0.4461
centimeters of mercury	kgs/sq meter	136.0
centimeters of mercury	pounds/sq ft	27.85
centimeters of mercury	pounds/sq in.	0.1934
centimeters/sec	feet/min	1.1969
centimeters/sec	feet/sec	0.03281
centimeters/sec	kilometers/hr	0.036
centimeters/sec	knots	0.1943
centimeters/sec	meters/min	0.6
centimeters/sec	miles/hr	0.02237
centimeters/sec	miles/min	3.728×10^{-4}
centimeters/sec/sec	feet/sec/sec	0.03281
centimeters/sec/sec	kms/hr/sec	0.036
centimeters/sec/sec	meters/sec/sec	0.01
centimeters/sec/sec	miles/hr/sec	0.02237
Chain	Inches	792.00
Chain	meters	20.12
Chains (surveyors' or Gunter's)	yards	22.00
circular mils	sq cms	5.067×10^{-6}
circular mils	sq mils	0.7854
Circumference	Radians	6.283
circular mils	sq inches	7.854×10^{-7}
Cords	cord feet	8
Cord feet	cu. feet	16
Coulomb	Statcoulombs	2.998×10^{9}
coulombs	faradays	1.036×10^{-5}
coulombs/sq cm	coulombs/sq in.	64.52
coulombs/sq cm	coulombs/sq meter	10^{4}
coulombs/sq in.	coulombs/sq cm	0.1550
coulombs/sq in.	coulombs/sq meter	1,550.
coulombs/sq meter	coulombs/sq cm	10^{-4}
coulombs/sq meter	coulombs/sq in.	6.452×10^{-4}
cubic centimeters	cu feet	3.531×10^{-5}
cubic centimeters	cu inches	0.06102

TO CONVERT	INTO	MULTIPLY BY
cubic centimeters	cu meters	10^{-6}
cubic centimeters	cu yards	1.308×10^{-6}
cubic centimeters	gallons (U. S. liq.)	2.642×10^{-4}
cubic centimeters	liters	0.001
cubic centimeters	pints (U.S. liq.)	2.113×10^{-3}
cubic centimeters	quarts (U.S. liq.)	1.057×10^{-3}
cubic feet	bushels (dry)	0.8036
cubic feet	cu cms	28,320.0
cubic feet	cu inches	1,728.0
cubic feet	cu meters	0.02832
cubic feet	cu yards	0.03704
cubic feet	gallons (U.S. liq.)	7.48052
cubic feet	liters	28.32
cubic feet	pints (U.S. liq.)	59.84
cubic feet	quarts (U.S. liq.)	29.92
cubic feet/min	cu cms/sec	472.0
cubic feet/min	gallons/sec	0.1247
cubic feet/min	liters/sec	0.4720
cubic feet/min	pounds of water/min	62.43
cubic feet/sec	million gals/day	0.646317
cubic feet/sec	gallons/min	448.831
cubic inches	cu cms	16.39
cubic inches	cu feet	5.787×10^{-4}
cubic inches	cu meters	1.639×10^{-5}
cubic inches	cu yards	2.143×10^{-5}
cubic inches	gallons	4.329×10^{-3}
cubic inches	liters	0.01639
cubic inches	mil-feet	1.061×10^{5}
cubic inches	pints (U.S. liq.)	0.03463
cubic inches	quarts (U.S. liq.)	0.01732
cubic meters	bushels (dry)	28.38
cubic meters	cu cms	10^{6}
cubic meters	cu feet	35.31
cubic meters	cu inches	61,023.0
cubic meters	cu yards	1.308
cubic meters	gallons (U.S. liq.)	264.2
cubic meters	liters	1,000.0
cubic meters	pints (U.S. liq.)	2,113.0
cubic meters	quarts (U.S. liq.)	1,057.
cubic yards	cu cms	7.646×10^{5}
cubic yards	cu feet	27.0
cubic yards	cu inches	46,656.0
cubic yards	cu meters	0.7646
cubic yards	gallons (U.S. liq.)	202.0
cubic yards	liters	764.6
cubic yards	pints (U.S. liq.)	1,615.9
cubic yards	quarts (U.S. liq.)	807.9
cubic yards/min	cubic ft/sec	0.45
cubic yards/min	gallons/sec	3.367
cubic yards/min	liters/sec	12.74

(Continued)

TO CONVERT	INTO	MULTIPLY BY
	D	
Dalton	Gram	1.650×10^{-24}
days	seconds	86,400.0
decigrams	grams	0.1
deciliters	liters	0.1
decimeters	meters	0.1
degrees (angle)	quadrants	0.01111
degrees (angle)	radians	0.01745
degrees (angle)	seconds	3,600.0
degrees/sec	radians/sec	0.01745
degrees/sec	revolutions/min	0.1667
degrees/sec	revolutions/sec	2.778×10^{-3}
dekagrams	grams	10.0
dekaliters	liters	10.0
dekameters	meters	10.0
Drams (apothecaries' or troy)	ounces (avoidupois)	0.1371429
Drams (apothecaries' or troy)	ounces (troy)	0.125
Drams (U.S., fluid or apoth.)	cubic cm.	3.6967
drams	grams	1.7718
drams	grains	27.3437
drams	ounces	0.0625
Dyne/cm	Erg/sq. millimeter	.01
Dyne/sq. cm.	Atmospheres	9.869×10^{-7}
Dyne/sq. cm.	Inch of Mercury at 0°C	2.953×10^{-5}
Dyne/sq. cm.	Inch of Water at 4°C	4.015×10^{-4}
dynes	grams	1.020×10^{-3}
dynes	joules/cm	10^{-7}
dynes	joules/meter (newtons)	10^{-5}
dynes	kilograms	1.020×10^{-6}
dynes	poundals	7.233×10^{-5}
dynes	pounds	2.248×10^{-6}
dynes/sq cm	bars	10^{-6}
	E	
Ell	Cm.	114.30
Ell	Inches	45
Em, Pica	Inch	.167
Em, Pica	Cm.	.4233
Erg/sec	Dyne – cm/sec	1.000
ergs	Btu	9.480×10^{-11}
ergs	dyne-centimeters	1.0
ergs	foot-pounds	7.367×10^{-8}
ergs	gram-calories	0.2389×10^{-7}
ergs	gram-cms	1.020×10^{-3}
ergs	horsepower-hrs	3.7250×10^{-14}
ergs	joules	10^{-7}

TO CONVERT	INTO	MULTIPLY BY
ergs	kg-calories	2.389×10^{-11}
ergs	kg-meters	1.020×10^{-8}
ergs	kilowatt-hrs	0.2778×10^{-13}
ergs	watt-hours	0.2778×10^{-10}
ergs/sec	Btu/min	$5,688 \times 10^{-9}$
ergs/sec	ft-lbs/min	4.427×10^{-6}
ergs/sec	ft-lbs/sec	7.3756×10^{-8}
ergs/sec	horsepower	1.341×10^{-10}
ergs/sec	kg-calories/min	1.433×10^{-9}
ergs/sec	kilowatts	10^{-10}
	F	
farads	microfarads	10^{6}
Faraday/sec	Ampere (absolute)	9.6500×10^{4}
faradays	ampere-hours	26.80
faradays	coulombs	9.649×10^{4}
Fathom	Meter	1.828804
fathoms	feet	6.0
feet	centimeters	30.48
feet	kilometers	3.048×10^{-4}
feet	meters	0.3048
feet	miles (naut.)	1.645×10^{-4}
feet	miles (stat.)	1.894×10^{-4}
feet	millimeters	304.8
feet	mils	1.2×10^{4}
feet of water	atmospheres	0.02950
feet of water	in. of mercury	0.8826
feet of water	kgs/sq cm	0.03048
feet of water	kgs/sq meter	304.8
feet of water	pounds/sq ft	62.43
feet of water	pounds/sq in.	0.4335
feet/min	cms/sec	0.5080
feet/min	feet/sec	0.01667
feet/min	kms/hr	0.01829
feet/min	meters/min	0.3048
feet/min	miles/hr	0.01136
feet/sec	cms/sec	30.48
feet/sec	kms/hr	1.097
feet/sec	knots	0.5921
feet/sec	meters/min	18.29
feet/sec	miles/hr	0.6818
feet/sec	miles/min	0.01136
feet/sec/sec	cms/sec/sec	30.48
feet/sec/sec	kms/hr/sec	1.097
feet/sec/sec	meters/sec/sec	0.3048
feet/sec/sec	miles/hr/sec	0.6818
feet/100 feet	per cent grade	1.0
Foot – candle	Lumen/sq. meter	10.764
foot-pounds	Btu	1.286×10^{-3}

(Continued)

TO CONVERT	INTO	MULTIPLY BY
foot-pounds	ergs	1.356×10^{7}
foot-pounds	gram-calories	0.3238
foot-pounds	hp-hrs	5.050×10^{-7}
foot-pounds	joules	1.356
foot-pounds	kg-calories	3.24×10^{-4}
foot-pounds	kg-meters	0.1383
foot-pounds	kilowatt-hrs	3.766×10^{-7}
foot-pounds/min	Btu/min	1.286×10^{-3}
foot-pounds/min	foot-pounds/sec	0.01667
foot-pounds/min	horsepower	3.030×10^{-5}
foot-pounds/min	kg-calories/min	3.24×10^{-4}
foot-pounds/min	kilowatts	2.260×10^{-5}
foot-pounds/sec	Btu/hr	4.6263
foot-pounds/sec	Btu/min	0.07717
foot-pounds/sec	horsepower	1.818×10^{-3}
foot-pounds/sec	kg-calories/min	0.01945
foot-pounds/sec	kilowatts	1.356×10^{-3}
Furlongs	miles (U.S.)	0.125
furlongs	rods	40.0
furlongs	feet	660.0
	G	
gallons	cu cms	3,785.0
gallons	cu feet	0.1337
gallons	cu inches	231.0
gallons	cu meters	3.785×10^{-3}
gallons	cu yards	4.951×10^{-3}
gallons	liters	3.785
gallons (liq. Br. Imp.)	gallons (U.S. liq.)	1.20095
gallons (U.S.)	gallons (Imp.)	0.83267
gallons of water	pounds of water	8.3453
gallons/min	cu ft/sec	2.228×10^{-3}
gallons/min	liters/sec	0.06308
gallons/min	cu ft/hr	8.0208
gausses	lines/sq in.	6.452
gausses	webers/sq cm	10^{-8}
gausses	webers/sq in.	6.452×10^{-8}
gausses	webers/sq meter	10^{-4}
gilberts	ampere-turns	0.7958
gilberts/cm	amp-turns/cm	0.7958
gilberts/cm	amp-turns/in	2.021
gilberts/cm	amp-turns/meter	79.58
Gills (British)	cubic cm.	142.07
gills	liters	0.1183
gills	pints (liq.)	0.25
Grade	Radian	.01571
Grains	drams (avoirdupois)	0.03657143
grains (troy)	grains (avdp)	1.0
grains (troy)	grams	0.06480

TO CONVERT	INTO	MULTIPLY BY
grains (troy)	ounces (avdp)	2.0833×10^{-3}
grains (troy)	pennyweight (troy)	0.04167
grains/U.S. gal	parts/million	17.118
grains/U.S. gal	pounds/million gal	142.86
grains/Imp. gal	parts/million	14.286
grams	dynes	980.7
grams	grains	15.43
grams	joules/cm	9.807×10^{-5}
grams	joules/meter (newtons)	9.807×10^{-3}
grams	kilograms	0.001
grams	milligrams	1,000.
grams	ounces (avdp)	0.03527
grams	ounces (troy)	0.03215
grams	poundals	0.07093
grams	pounds	2.205×10^{-3}
grams/cm	pounds/inch	5.600×10^{-3}
grams/cu cm	pounds/cu ft	62.43
grams/cu cm	pounds/cu in	0.03613
grams/cu cm	pounds/mil-foot	3.405×10^{-7}
grams/liter	grains/gal	58.417
grams/liter	pounds/ gal	8.345
grams/liter	pounds/cu ft	0.062427
grams/liter	parts/million	1,000.0
grams/sq cm	pounds/sq ft	2.0481
gram-calories	Btu	3.9683×10^{-3}
gram-calories	ergs	4.1868×10^{7}
gram-calories	foot-pounds	3.0880
gram-calories	horsepower-hrs	1.5596×10^{-6}
gram-calories	kilowatt-hrs	1.1630×10^{-6}
gram-calories	watt-hrs	1.1630×10^{-3}
gram-calories/sec	Btu/hr	14.286
gram-centimeters	Btu	9.297×10^{-8}
gram-centimeters	ergs	980.7
gram-centimeters	joules	9.807×10^{-5}
gram-centimeters	kg-cal	2.343×10^{-8}
gram-centimeters	kg-meters	10^{-5}
	H	
Hand	Cm.	10.16
hectares	acres	2.471
hectares	sq feet	1.076×10^{5}
hectograms	grams	100.0
hectoliters	liters	100.0
hectometers	meters	100.0
hectowatts	watts	100.0
henries	millihenries	1,000.0
Hogsheads (British)	cubic ft.	10.114
Hogsheads (U.S.)	cubic ft.	8.42184
Hogsheads (U.S.)	gallons (U.S.)	63

(Continued)

TO CONVERT	INTO	MULTIPLY BY
horsepower	Btu/min	42.44
horsepower	foot-lbs/min	33,000.
horsepower	foot-lbs/sec	550.0
horsepower (metric) (542.5 ft lb/sec)	horsepower (550 ft lb/sec)	0.9863
horsepower (550 ft lb/sec)	horsepower (metric) (542.5 ft lb/sec)	1.014
horsepower	kg-calories/min	10.68
horsepower	kilowatts	0.7457
horsepower	watts	745.7
horsepower (boiler)	Btu/hr	33.479
horsepower (boiler)	kilowatts	9.803
horsepower-hrs	Btu	2,547.
horsepower-hrs	ergs	2.6845×10^{13}
horsepower-hrs	foot-lbs	1.98×10^{6}
horsepower-hrs	gram-calories	641,190.
horsepower-hrs	joules	2.684×10^{6}
horsepower-hrs	kg-calories	641.1
horsepower-hrs	kg-meters	2.737×10^{5}
horsepower-hrs	kilowatt-hrs	0.7457
hours	days	4.167×10^{-2}
hours	weeks	5.952×10^{-3}
Hundredweights (long)	pounds	112
Hundredweights (long)	tons (long)	0.05
Hundredweights (short)	ounces (avoirdupois)	1600
Hundredweights (short)	pounds	100
Hundredweights (short)	tons (metric)	0.0453592
Hundredweights (short)	tons (long)	0.0446429
I		
inches	centimeters	2.540
inches	meters	2.540×10^{-2}
inches	miles	1.578×10^{-5}
inches	millimeters	25.40
inches	mils	1,000.0
inches	yards	2.778×10^{-2}
inches of mercury	atmospheres	0.03342
inches of mercury	feet of water	1.133
inches of mercury	kgs/sq cm	0.03453
inches of mercury	kgs/sq meter	345.3
inches of mercury	pounds/sq ft	70.73
inches of mercury	pounds/sq in.	0.4912
inches of water (at 4°C)	atmospheres	2.458×10^{-3}
inches of water (at 4°C)	inches of mercury	0.07355
inches of water (at 4°C)	kgs/sq cm	2.540×10^{-3}
inches of water (at 4°C)	ounces/sq in.	0.5781
inches of water (at 4°C)	pounds/sq ft	5.204
inches of water (at 4°C)	pounds/sq in.	0.03613
International Ampere	Ampere (absolute)	.9998
International Volt	Volts (absolute)	1.0003

TO CONVERT	INTO	MULTIPLY BY
International volt	Joules (absolute)	1-593 x 10^{-19}
International volt	Joules	9.654 x 10^{4}
	J	
joules	Btu	9.480 x 10^{-4}
joules	ergs	10^{7}
joules	foot-pounds	0.7376
joules	kg-calories	2.389 x 10^{-4}
joules	kg-meters	0.1020
joules	watt-hrs	2.778 x 10^{-4}
joules/cm	grams	1.020 x 10^{4}
joules/cm	dynes	10^{7}
joules/cm	joules/meter (newtons)	100.0
joules/cm	poundals	723.3
joules/cm	pounds	22.48
	K	
kilograms	dynes	980,665.
kilograms	grams	1,000.0
kilograms	joules/cm	0.09807
kilograms	joules/meter (newtons)	9.807
kilograms	poundals	70.93
kilograms	pounds	2.205
kilograms	tons (long)	9.842 x 10^{-4}
kilograms	tons (short)	1.102 x 10^{-3}
kilograms/cu meter	grams/cu cm	0.001
kilograms/cu meter	pounds/cu ft	0.06243
kilograms/cu meter	pounds/cu in.	3.613 x 10^{-5}
kilograms/cu meter	pounds/mil-foot	3.405 x 10^{-10}
kilograms/meter	pounds/ft	0.6720
Kilogram/sq. cm.	Dynes	980,665
kilograms/sq cm	atmospheres	0.9678
kilograms/sq cm	feet of water	32.81
kilograms/sq cm	inches of mercury	28.96
kilograms/sq cm	pounds/sq ft	2,048.
kilograms/sq cm	pounds/sq in.	14.22
kilograms/sq meter	atmospheres	9.678 x 10^{-5}
kilograms/sq meter	bars	98.07 x 10^{-6}
kilograms/sq meter	feet of water	3.281 x 10^{-3}
kilograms/sq meter	inches of mercury	2.896 x 10^{-3}
kilograms/sq meter	pounds/sq ft	0.2048
kilograms/sq meter	pounds/sq in.	1.422 x 10^{-3}
kilograms/sq mm	kgs/sq meter	10^{6}
kilogram-calories	Btu	3.968
kilogram-calories	foot-pounds	3,088.
kilogram-calories	hp-hrs	1.560 x 10^{-3}
kilogram-calories	joules	4,186.
kilogram-calories	kg-meters	426.9

(Continued)

TO CONVERT	INTO	MULTIPLY BY
kilogram-calories	kilojoules	4.186
kilogram-calories	kilowatt-hrs	1.163×10^{-3}
kilogram meters	Btu	9.294×10^{-3}
kilogram meters	ergs	9.804×10^{7}
kilogram meters	foot-pounds	7.233
kilogram meters	joules	9.804
kilogram meters	kg-calories	2.342×10^{-3}
kilogram meters	kilowatt-hrs	2.723×10^{-6}
kilolines	maxwells	1,000.0
kiloliters	liters	1,000.0
kilometers	centimeters	10^{5}
kilometers	feet	3,281.
kilometers	inches	3.937×10^{4}
kilometers	meters	1,000.0
kilometers	miles	0.6214
kilometers	millimeters	10^{6}
kilometers	yards	1,094.
kilometers/hr	cms/sec	27.78
kilometers/hr	feet/min	54.68
kilometers/hr	feet/sec	0.9113
kilometers/hr	knots	0.5396
kilometers/hr	meters/min	16.67
kilometers/hr	miles/hr	0.6214
kilometers/hr/sec	cms/sec/sec	27.78
kilometers/hr/sec	ft/sec/sec	0.9113
kilometers/hr/sec	meters/sec/sec	0.2778
kilometers/hr/sec	miles/hr/sec	0.6214
kilowatts	Btu/min	56.92
kilowatts	foot-lbs/min	4.426×10^{4}
kilowatts	foot-lbs/sec	737.6
kilowatts	horsepower	1.341
kilowatts	kg-calories/min	14.34
kilowatts	watts	1,000.0
kilowatt-hrs	Btu	3,413.
kilowatt-hrs	ergs	3.600×10^{13}
kilowatt-hrs	foot-lbs	2.655×10^{6}
kilowatt-hrs	gram-calories	859,850.
kilowatt-hrs	horsepower-hrs	1.341
kilowatt-hrs	joules	3.6×10^{6}
kilowatt-hrs	kg-calories	860.5
kilowatt-hrs	kg-meters	3.671×10^{5}
kilowatt-hrs	pounds of water evaporated from and at 212° F.	3.53
kilowatt-hrs	pounds of water raised from 62° to 212° F.	22.75
knots	feet/hr	6,080.

TO CONVERT	INTO	MULTIPLY BY
knots	kilometers/hr	1.8532
knots	nautical miles/hr	1.0
knots	statute miles/hr	1.151
knots	yards/hr	2,027.
knots	feet/sec	1.689
	L	
league	miles (approx.)	3.0
Light year	Miles	5.9×10^{12}
Light year	Kilometers	9.46091×10^{12}
lines/sq cm	gausses	1.0
lines/sq in.	gausses	0.1550
lines/sq in.	webers/sq cm	1.550×10^{-9}
lines/sq in.	webers/sq in.	10^{-8}
lines/sq in.	webers/sq meter	1.550×10^{-5}
links (engineer's)	inches	12.0
links (surveyor's)	inches	7.92
liters	bushels (U.S. dry)	0.02838
liters	cu cm	1,000.0
liters	cu feet	0.03531
liters	cu inches	61.02
liters	cu meters	0.001
liters	cu yards	1.308×10^{-3}
liters	gallons (U.S. liq.)	0.2642
liters	pints (U.S. liq.)	2.113
liters	quarts (U.S. liq.)	1.057
liters/min	cu ft/sec	5.886×10^{-4}
liters/min	gals/sec	4.403×10^{-3}
lumens/sq ft	foot-candles	1.0
Lumen	Spherical candle power	.07958
Lumen	Watt	.001496
Lumen/sq. ft.	Lumen/sq. meter	10.76
lux	foot-candles	0.0929
	M	
maxwells	kilolines	0.001
maxwells	webers	10^{-8}
megalines	maxwells	10^{6}
megohms	microhms	10^{12}
megohms	ohms	10^{6}
meters	centimeters	100.0
meters	feet	3.281
meters	inches	39.37
meters	kilometers	0.001
meters	miles (naut.)	5.396×10^{-4}
meters	miles (stat.)	6.214×10^{-4}
meters	millimeters	1,000.0
meters	yards	1.094
meters	varas	1.179

(Continued)

TO CONVERT	INTO	MULTIPLY BY
meters/min	cms/sec	1.667
meters/min	feet/min	3.281
meters/min	feet/sec	0.05468
meters/min	kms/hr	0.06
meters/min	knots	0.03238
meters/min	miles/hr	0.03728
meters/sec	feet/min	196.8
meters/sec	feet/sec	3.281
meters/sec	kilometers/hr	3.6
meters/sec	kilometers/min	0.06
meters/sec	miles/hr	2.237
meters/sec	miles/min	0.03728
meters/sec/sec	cms/sec/sec	100.0
meters/sec/sec	ft/sec/sec	3.281
meters/sec/sec	kms/hr/sec	3.6
meters/sec/sec	miles/hr/sec	2.237
meter-kilograms	cm-dynes	9.807×10^{7}
meter-kilograms	cm-grams	10^{5}
meter-kilograms	pound-feet	7.233
microfarad	farads	10^{-6}
micrograms	grams	10^{-6}
microhms	megohms	10^{-12}
microhms	ohms	10^{-6}
microliters	liters	10^{-6}
Microns	meters	1×10^{-6}
miles (naut.)	feet	6,080.27
miles (naut.)	kilometers	1.853
miles (naut.)	meters	1,853.
miles (naut.)	miles (statute)	1.1516
miles (naut.)	yards	2,027.
miles (statute)	centimeters	1.609×10^{5}
miles (statute)	feet	5,280.
miles (statute)	inches	6.336×10^{4}
miles (statute)	kilometers	1.609
miles (statute)	meters	1,609.
miles (statute)	miles (naut.)	0.8684
miles (statute)	yards	1,760.
miles/hr	cms/sec	44.70
miles/hr	feet/min	88.
miles/hr	feet/sec	1.467
miles/hr	kms/hr	1.609
miles/hr	kms/min	0.02682
miles/hr	knots	0.8684
miles/hr	meters/min	26.82
miles/hr	miles/min	0.1667
miles/hr/sec	cms/sec/sec	44.70
miles/hr/sec	feet/sec/sec	1.467
miles/hr/sec	kms/hr/sec	1.609
miles/hr/sec	meters/sec/sec	0.4470
miles/min	cms/sec	2,682.
miles/min	feet/sec	88.

TO CONVERT	INTO	MULTIPLY BY
miles/min	kms/min	1.609
miles/min	knots/min	0.8684
miles/min	miles/hr	60.0
mil-feet	cu inches	9.425×10^{-6}
milliers	kilograms	1,000.
Millimicrons	meters	1×10^{-9}
Milligrams	grains	0.01543236
milligrams	grams	0.001
milligrams/liter	parts/million	1.0
millihenries	henries	0.001
milliliters	liters	0.001
millimeters	centimeters	0.1
millimeters	feet	3.281×10^{-3}
millimeters	inches	0.03937
millimeters	kilometers	10^{-6}
millimeters	meters	0.001
millimeters	miles	6.214×10^{-7}
millimeters	mils	39.37
millimeters	yards	1.094×10^{-3}
million gals/day	cu ft/sec	1.54723
mils	centimeters	2.540×10^{-3}
mils	feet	8.333×10^{-5}
mils	inches	0.001
mils	kilometers	2.540×10^{-8}
mils	yards	2.778×10^{-5}
miner's inches	cu ft/min	1.5
Minims (British)	cubic cm.	0.059192
Minims (U.S., fluid)	cubic cm.	0.061612
minutes (angles)	degrees	0.01667
minutes (angles)	quadrants	1.852×10^{-4}
minutes (angles)	radians	2.909×10^{-4}
minutes (angles)	seconds	60.0
myriagrams	kilograms	10.0
myriameters	kilometers	10.0
myriawatts	kilowatts	10.0
	N	
nepers	decibels	8.686
Newton	Dynes	1 x 105
	O	
OHM (International)	OHM (absolute)	1.0005
ohms	megohms	10^{-6}
ohms	microhms	10^{6}
ounces	drams	16.0
ounces	grains	437.5
ounces	grams	28.349527
ounces	pounds	0.0625

(Continued)

TO CONVERT	INTO	MULTIPLY BY
ounces	ounces (troy)	0.9115
ounces	tons (long)	2.790×10^{-5}
ounces	tons (metric)	2.835×10^{-5}
ounces (fluid)	cu inches	1.805
ounces (fluid)	liters	0.02957
ounces (troy)	grains	480.0
ounces (troy)	grams	31.103481
ounces (troy)	ounces (avdp.)	1.09714
ounces (troy)	pennyweights (troy)	20.0
ounces (troy)	pounds (troy)	0.08333
Ounce/sq. inch	Dynes/sq. cm.	4309
ounces/sq in.	pounds/sq in.	0.0625
	P	
Parsec	Miles	19×10^{12}
Parsec	Kilometers	3.084×10^{13}
parts/million	grains/U.S. gal	0.0584
parts/million	grains/Imp. gal	0.07016
parts/million	pounds/million gal	8.345
Pecks (British)	cubic inches	554.6
Pecks (British)	liters	9.091901
Pecks (U.S.)	bushels	0.25
Pecks (U.S.)	cubic inches	537.605
Pecks (U.S.)	liters	8.809582
Pecks (U.S.)	quarts (dry)	8
pennyweights (troy)	grains	24.0
pennyweights (troy)	ounces (troy)	0.05
pennyweights (troy)	grams	1.55517
pennyweights (troy)	pounds (troy)	4.1667×10^{-3}
pints (dry)	cu inches	33.60
pints (liq.)	cu cms.	473.2
pints (liq.)	cu feet	0.01671
pints (liq.)	cu inches	28.87
pints (liq.)	cu meters	4.732×10^{-4}
pints (liq.)	cu yards	6.189×10^{-4}
pints (liq.)	gallons	0.125
pints (liq.)	liters	0.4732
pints (liq.)	quarts (liq.)	0.5
Planck's quantum	Erg – second	6.624×10^{-27}
Poise	Gram/cm. sec.	1.00
Pounds (avoirdupois)	ounces (troy)	14.5833
poundals	dynes	13,826.
poundals	grams	14.10
poundals	joules/cm	1.383×10^{-3}
poundals	joules/meter (newtons)	0.1383
poundals	kilograms	0.01410
poundals	pounds	0.03108
pounds	drams	256.
pounds	dynes	44.4823×10^{4}
pounds	grains	7,000.

TO CONVERT	INTO	MULTIPLY BY
pounds	grams	453.5924
pounds	joules/cm	0.04448
pounds	joules/meter (newtons)	4.448
pounds	kilograms	0.4536
pounds	ounces	16.0
pounds	ounces (troy)	14.5833
pounds	poundals	32.17
pounds	pounds (troy)	1.21528
pounds	tons (short)	0.0005
pounds (troy)	grains	5,760.
pounds (troy)	grams	373.24177
pounds (troy)	ounces (avdp.)	13.1657
pounds (troy)	ounces (troy)	12.0
pounds (troy)	pennyweights (troy)	240.0
pounds (troy)	pounds (avdp.)	0.822857
pounds (troy)	tons (long)	3.6735×10^{-4}
pounds (troy)	tons (metric)	3.7324×10^{-4}
pounds (troy)	tons (short)	4.1143×10^{-4}
pounds of water	cu feet	0.01602
pounds of water	cu inches	27.68
pounds of water	gallons	0.1198
pounds of water/min	cu ft/sec	2.670×10^{-4}
pound-feet	cm-dynes	1.356×10^{7}
pound-feet	cm-grams	13,825.
pound-feet	meter-kgs	0.1383
pounds/cu ft	grams/cu cm	0.01602
pounds/cu ft	kgs/cu meter	16.02
pounds/cu ft	pounds/cu in.	5.787×10^{-4}
pounds/cu ft	pounds/mil-foot	5.456×10^{-9}
pounds/cu in.	gms/cu cm	27.68
pounds/cu in.	kgs/cu meter	2.768×10^{4}
pounds/cu in.	pounds/cu ft	1,728.
pounds/cu in.	pounds/mil-foot	9.425×10^{-6}
pounds/ft	kgs/meter	1.488
pounds/in.	gms/cm	178.6
pounds/mil-foot	gms/cu cm	2.306×10^{6}
pounds/sq ft	atmospheres	4.725×10^{-4}
pounds/sq ft	feet of water	0.01602
pounds/sq ft	inches of mercury	0.01414
pounds/sq ft	kgs/sq meter	4.882
pounds/sq ft	pounds/sq in.	6.944×10^{-3}
pounds/sq in.	atmospheres	0.06804
pounds/sq in.	feet of water	2.307
pounds/sq in.	inches of mercury	2.036
pounds/sq in.	kgs/sq meter	703.1
pounds/sq in.	pounds/sq ft	144.0

Q

TO CONVERT	INTO	MULTIPLY BY
quadrants (angle)	degrees	90.0
quadrants (angle)	minutes	5,400.0

(Continued)

TO CONVERT	INTO	MULTIPLY BY
quadrants (angle)	radians	1.571
quadrants (angle)	seconds	3.24×10^5
quarts (dry)	cu inches	67.20
quarts (liq.)	cu cms	946.4
quarts (liq.)	cu feet	0.03342
quarts (liq.)	cu inches	57.75
quarts (liq.)	cu meters	9.464×10^{-4}
quarts (liq.)	cu yards	1.238×10^{-3}
quarts (liq.)	gallons	0.25
quarts (liq.)	liters	0.9463
	R	
radians	degrees	57.30
radians	minutes	3,438.
radians	quadrants	0.6366
radians	seconds	2.063×10^5
radians/sec	degrees/sec	57.30
radians/sec	revolutions/min	9.549
radians/sec	revolutions/sec	0.1592
radians/sec/sec	revs/min/min	573.0
radians/sec/sec	revs/min/sec	9.549
radians/sec/sec	revs/sec/sec	0.1592
revolutions	degrees	360.0
revolutions	quadrants	4.0
revolutions	radians	6.283
revolutions/min	degrees/sec	6.0
revolutions/min	radians/sec	0.1047
revolutions/min	revs/sec	0.01667
revolutions/min/min	radians/sec/sec	1.745×10^{-3}
revolutions/min/min	revs/min/sec	0.01667
revolutions/min/min	revs/sec/sec	2.778×10^{-4}
revolutions/sec	degrees/sec	360.0
revolutions/sec	radians/sec	6.283
revolutions/sec	revs/min	60.0
revolutions/sec/sec	radians/sec/sec	6.283
revolutions/sec/sec	revs/min/min	3,600.0
revolutions/sec/sec	revs/min/sec	60.0
Rod	Chain (Gunters)	.25
Rod	Meters	5.029
Rods (Surveyors' meas.)	yards	5.5
rods	feet	16.5
	S	
Scruples	grains	20
seconds (angle)	degrees	2.778×10^{-4}
seconds (angle)	minutes	0.01667
seconds (angle)	quadrants	3.087×10^{-6}
seconds (angle)	radians	4.848×10^{-6}
Slug	Kilogram	14.59

TO CONVERT	INTO	MULTIPLY BY
Slug	Pounds	32.17
Sphere	Steradians	12.57
square centimeters	circular mils	1.973×10^{5}
square centimeters	sq feet	1.076×10^{-3}
square centimeters	sq inches	0.1550
square centimeters	sq meters	0.0001
square centimeters	sq miles	3.861×10^{-11}
square centimeters	sq millimeters	100.0
square centimeters	sq yards	1.196×10^{-4}
square feet	acres	2.296×10^{-5}
square feet	circular mils	1.833×10^{8}
square feet	sq cms	929.0
square feet	sq inches	144.0
square feet	sq meters	0.09290
square feet	sq miles	3.587×10^{-8}
square feet	sq millimeters	9.290×10^{4}
square feet	sq yards	0.1111
square inches	circular mils	1.273×10^{6}
square inches	sq cms	6.452
square inches	sq feet	6.944×10^{-3}
square inches	sq millimeters	645.2
square inches	sq mils	10^{6}
square inches	sq yards	7.716×10^{-4}
square kilometers	acres	247.1
square kilometers	sq cms	10^{10}
square kilometers	sq ft	10.76×10^{6}
square kilometers	sq inches	1.550×10^{9}
square kilometers	sq meters	10^{6}
square kilometers	sq miles	0.3861
square kilometers	sq yards	1.196×10^{6}
square meters	acres	2.471×10^{-4}
square meters	sq cms	10^{4}
square meters	sq feet	10.76
square meters	sq inches	1,550.
square meters	sq miles	3.861×10^{-7}
square meters	sq millimeters	10^{6}
square meters	sq yards	1.196
square miles	acres	640.0
square miles	sq feet	27.88×10^{6}
square miles	sq kms	2.590
square miles	sq meters	2.590×10^{6}
square miles	sq yards	3.098×10^{6}
square millimeters	circular mils	1,973.
square millimeters	sq cms	0.01
square millimeters	sq feet	1.076×10^{-5}
square millimeters	sq inches	1.550×10^{-3}
square mils	circular mils	1.273
square mils	sq cms	6.452×10^{-6}
square mils	sq inches	10^{-6}
square yards	acres	2.066×10^{-4}

(Continued)

TO CONVERT	INTO	MULTIPLY BY
square yards	sq cms	8,361.
square yards	sq feet	9.0
square yards	sq inches	1,296.
square yards	sq meters	0.8361
square yards	sq miles	3.228×10^{-7}
square yards	sq millimeters	8.361×10^{5}
T		
temperature (°C) +273	absolute temperature (°C)	1.0
temperature (°C) +17.78	temperature (°F)	1.8
temperature (°F) +460	absolute temperature (°F)	1.0
temperature (°F) −32	temperature (°C)	5/9
tons (long)	kilograms	1,016.
tons (long)	pounds	2,240.
tons (long)	tons (short)	1.120
tons (metric)	kilograms	1,000.
tons (metric)	pounds	2,205.
tons (short)	kilograms	907.1848
tons (short)	ounces	32,000.
tons (short)	ounces (troy)	29,166.66
tons (short)	pounds	2,000.
tons (short)	pounds (troy)	2,430.56
tons (short)	tons (long)	0.89287
tons (short)	tons (metric)	0.9078
tons (short)/sq ft	kgs/sq meter	9,765.
tons (short)/sq ft	pounds/sq in.	2,000.
tons of water/24 hrs	pounds of water/hr	83.333
tons of water/24 hrs	gallons/min	0.16643
tons of water/24 hrs	cu ft/hr	1.3349
V		
Volt/inch	Volt/cm.	.39370
Volt (absolute)	Statvolts	.003336
W		
watts	Btu/hr	3.4129
watts	Btu/min	0.05688
watts	ergs/sec	107.
watts	foot-lbs/min	44.27
watts	foot-lbs/sec	0.7378
watts	horsepower	1.341×10^{-3}

TO CONVERT	INTO	MULTIPLY BY
watts	horsepower (metric)	1.360×10^{-3}
watts	kg-calories/min	0.01433
watts	kilowatts	0.001
Watts (Abs.)	B.T.U. (mean)/min.	0.056884
Watts (Abs.)	joules/sec.	1
watt-hours	Btu	3.413
watt-hours	ergs	3.60×10^{10}
watt-hours	foot-pounds	2,656.
watt-hours	gram-calories	859.85
watt-hours	horsepower-hrs	1.341×10^{-3}
watt-hours	kilogram-calories	0.8605
watt-hours	kilogram-meters	367.2
watt-hours	kilowatt-hrs	0.001
Watt (International)	Watt (absolute)	1.0002
webers	maxwells	10^8
webers	kilolines	10^5

References

1. American Petroleum Institute, Refinery Practice 520, 1969, pp. 8.
2. American Petroleum Institute, Refinery Practice 520, Part I, 1973, pp. 18.
3. American Petroleum Institute, Refinery Practice 520, Part I, 1973, pp. 26.
4. Lapple, C. E., "Isothermal and Adiabatic Flow of Compressible Fluids," Trans. AIChE, 39, 385 (1943).
5. Loeb, M. B., "Graphical Solution of Compressible Fluid Flow Problems," Report TR-256-D, 1965, John F. Kennedy Space Center.
6. Moody, L. F., Trans. ASME 66 (1944).
7. American Petroleum Institute, Refinery Practice 621, 1969, pp. 45.
8. Tau, S. H., "Flare Design Simplified," *Hydrocarbon Processing,* "Waste Treatment Flare Stack Design Handbook," Houston, Texas, 1968.
9. Perry, Robert H., Chilton, Cecil H., *Chemical Engineer's Handbook,* Fifth Edition, 1972, McGraw-Hill, NY.

10. American Petroleum Institute, Refinery Practice 521, 1969, pp. 50.
11. Kent, G. B., "Practical Design of Flare Stacks," *Hydrocarbon Processing,* Houston, Texas, 1968.
12. American Petroleum Institute, Refinery practice 521, 1969, pp. 35.
13. American Petroleum Institute, Refinery practice 521, 1969, pp. 64.
14. American Petroleum Institute's Manual on "Disposal of Refinery Wastes," Vol. II, Chapter 9, 1976, pp. 53.
15. Cheremisinoff, P. N. and Young, R. A., *Pollution Engineering Practice Handbook,* Ann Arbor Science Pub., Ann Arbor, MI (1976).
16. Cheremisinoff, P. N. and Young, R. A., editors, *Air Pollution Control and Design Handbook,* Part 1, Marcel Dekker Inc., NY (1977).
17. American Petroleum Institute, Refinery practice, 521, 1969, pp. 67.
18. American Petroleum Institute, Refinery practice, 521, 1969, pp. 54.
19. Azbel, D. and Cheremisinoff, N. P., *Chemical and Process Equipment Design,* Ann Arbor Science Pub., Ann Arbor, MI (1983).
20. American Petroleum Institute, Refinery practice 520, Part II, 1973, pp. 31.

INDEX